THE HUMANIST
PRINCIPLE

A pioneer of cardiovascular medicine, and the author of more than 400 scientific and medical publications, **Felix Unger** (1946–) is one of the world's foremost heart specialists. He is the inventor of the ellipsoid heart, an artificial heart that enabled him to perform the first clinical artificial heart implantation in Europe. Dr Unger graduated in 1971 from the University of Vienna and is currently President of both the Academy of Sciences and Arts and the European Institute of Health.

Daisaku Ikeda (1928–) is President of Soka Gakkai International, a Buddhist network that actively promotes peace, culture and education, whose members come from over 190 countries worldwide. He is the author of more than 100 books on Buddhist themes, and he received the United Nations Peace Award in 1983. Ikeda has also been the recipient of numerous academic honours from research and academic institutions in the UK, China, Brazil, USA and other countries.

THE HUMANIST PRINCIPLE

On Compassion and Tolerance

Felix Unger
and
Daisaku Ikeda

I.B. TAURIS
LONDON · NEW YORK

Published in 2016 by
I.B.Tauris & Co. Ltd
London • New York
www.ibtauris.com

Original copyright © 2016 Felix Unger and Daisaku Ikeda
English translation copyright © 2016 Soka Gakkai

References to websites were correct at the time of writing.

ISBN (HB): 978 1 78453 782 1
ISBN (PB): 978 1 78453 783 8
eISBN: 978 1 78672 155 6
ePDF: 978 1 78673 155 5

A full CIP record for this book is available from the British Library
A full CIP record is available from the Library of Congress

Library of Congress Catalog Card Number: available

Typeset in Garamond by Initial Typesetting Services, Edinburgh

Contents

The European Academy of Sciences and Arts and SGI

Ikeda: Human existence today is threatened by both direct and structural violence. Is it the hard power of military and economic might or the soft power of dialogue that will bring peace, security and coexistence in the twenty-first century? The choice between the two has been sharply debated since the terrorist attacks of 11 September 2001. Has the globalization of dialogue kept pace with the rapid systemization and globalization of the economy and of communications? In July 1997, at Soka University in Tokyo, you made a highly instructive comment: 'Our ability to engage in dialogue will determine the further fate of this Earth'.

Never before have sincere intercultural and interreligious dialogues been as necessary as they are now. I shall be supremely happy if this dialogue with you, as president of an institute of leading thinkers, the European Academy of Sciences and Arts, contributes to the globalization of dialogue.

Unger: To that end, I intend to give my best to the project.

Today, owing to a growing emphasis on materialistic values, many religions are losing sight of the traditional values that they once held in common. The globalization you mention gives impetus to this process. As the value put on life diminishes, deadly forms of violence become more common. On the one hand, religion declines; on the other, materialism grows more prevalent. Halting this trend requires us to recall values inherited from the distant past.

Ikeda: I see what you mean. As Tolstoy observed, recognition of the sacredness of every individual life is the first and only basis of all morality.[1] Christianity, Buddhism and other religions share fundamental ethical prohibitions against killing. Though the terminology may differ, non-violence and the injunction against killing are such prohibitions. To make the twenty-first century a century free of violence and killing we must make those prohibitions the foundation of global ethics. This is why interfaith dialogues are so necessary. In your speech at Soka University you said, 'Interreligious dialogues must form the axis of all intercultural dialogue and a foundation for the creation of culture on a global scale'.

Unger: Soka Gakkai International (SGI), with which I have had repeated contacts for more than ten years, is a powerful Buddhist organization seeking to define reasonable goals for humanity from the Buddhist tradition and apply them to the real world.

This tradition is exemplified by you, Mr Ikeda. Within the frame of extensive theological, organizational and publication capacities, you strive to express the value of humanity in our world. Through international cultural exchanges, you have transmitted the message of peace with startling force. As we begin this dialogue, I would like to ask what has motivated you personally.

Ikeda: My own experience of war has been an important source of motivation for my work for peace. During World War II, the evil of ultranationalism shattered the peace of our family. Our home was destroyed in air raids, and my oldest brother died at the front. Although my mother was a courageous woman, she was crushed by the news of his death. Later, when his ashes were returned to us, she sat holding the container, her shoulders shaking in grief. I shall always remember the sight. At an early age, I learned that, no matter how reality may be papered over with lies, war is foolish, hideous misery.

A second and later source of motivation was the spirit I inherited from my mentors in life. In the very midst of World War II, Tsunesaburo Makiguchi (1871–1944), first president of Soka Gakkai, and his protégé Josei Toda (1900–58) – my direct mentor – strove to uphold in action the philosophy of the Japanese Buddhist leader Nichiren (1222–82), who emphasized the dignity of life. The militarist authorities threw Makiguchi in prison, where he died. Toda, too, was imprisoned but survived to carry on Makiguchi's legacy in the struggle for peace. I consider myself Toda's spiritual heir. His deep wish was to rid the world of misery. Working to realize his goal is my whole life.

A third impetus for my work is my sense of social mission as a person of religion. Whether direct or structural, violence causes great suffering. Instead of merely contemplating others' suffering, we must rise up and practise the Buddhist doctrine of 'removing suffering and giving joy'. Sympathetic action of this kind constitutes the whole spirit of Mahayana Buddhism. Nichiren, whom we of Soka Gakkai revere, taught this spirit in his treatise 'On Establishing the Correct Teaching for the Peace of the Land'. One cannot find personal salvation if one selfishly ignores the misery imposed on others by the threat of violence. War is the most inhuman form of such threat.

Unger: The founders of Soka Gakkai, particularly second president Josei Toda, worked actively for peace during World War II and incurred personal suffering and loss of freedom. This has been the starting point for Soka Gakkai's earnest engagement for peace and a world free of nuclear weapons. I have the greatest respect for that work.

Ikeda: Toda's resolution to prohibit nuclear arms and his insistence that anything jeopardizing people's inviolable right to live is a monstrous evil constituted his testament to me and my young peers. Nuclear weapons have the power to destroy all life – the human race and the whole ecological system. We must make all humanity aware of this evil. Part of our work to this end are the exhibitions entitled 'Nuclear Arms: Threat to Humanity', and, 'Nuclear Arms – Threat to Our World', which we have exhibited in twenty-four nations and thirty-nine cities, including China, the former Soviet Union, the United Nations Headquarters in New York, and Vienna, the capital of your homeland. In promoting cultural exchanges, we have held several other exhibitions in Vienna, from 'Maki-e Lacquer and Oriental Ceramics', 'Treasures of Japanese Art' and 'The Lotus Sutra and Its World: Buddhist Manuscripts of the Great Silk Road' to 'Dialogue with Nature', an exhibition of my own photographs that was shown twice in Vienna and also at the Hungarian National Gallery, as a result – to my honour – of support extended by your academy. I would like to take this opportunity to express my gratitude again.

Unger: Though I do not know your intentions precisely, this is how I interpret your photographic goals. Man is an element of nature and is on the right road only when grounded in nature. In your landscape photographs, you reveal this symbolically in a fascinating way – pristine nature, cultivated nature, and completely manmade gardens. Your images show not only the possibility of peaceful coexistence with nature, but also the foundation in nature for peaceful human togetherness.

In spite of this possibility, however, man is capable of dominating nature and destroying the basis of life. In your book *LIFE: An*

Enigma, a Precious Jewel, you explain this from the standpoint of Buddhist tradition.

Ikeda: Yes. In that book I examine relations between nature and humanity. In the Buddhist view, the fundamental Law of the universe manifests in all individual life forms – humanity, non-human nature, and the stars of the firmament. Individual life forms are not isolated parts but are integrated with the cosmic life force. Put another way, the part is the whole, and the whole is the part. Human beings and non-human nature are integrated parts of the same cosmic life force. They are unique in the individual sense and whole in the symbiotic, cosmic sense; irrevocably integrated with each other. Therefore, to destroy nature is to destroy humanity. Buddhism cannot countenance human destruction of nature or its exploitation to satisfy selfish greed. That is why Buddhists regard environmental devastation, social devastation in the form of rampant violence and extreme poverty, as well as spiritual devastation in the form of hypertrophic egoism, as all sharing a root cause.

The European Academy of Sciences and Arts, which you founded with the support of European leaders and intellectuals, shares the SGI's outlook on various issues. By means of intercultural and interfaith dialogues the academy attempts to restructure relationships between human beings and the spirit, society, and nature. I am proud to participate in the academy's noble mission as an honorary senator.

Unger: You play an active part by conducting dialogues with people from all over the world. The position of honorary senator is conferred by our academy on people of action like you. In my professional work as a cardiac surgeon, I understand the importance of decisiveness, resolution, and action unimpeded by lengthy philosophical reflection.

Parental Influences

Ikeda: Now permit me to ask you a few questions. You are the president of the European Academy of Sciences and Arts, a renowned cardiac surgeon and a distinguished thinker. I imagine that certain people who instructed you either in your field of learning or in private life have earned your profound respect. Who are some of them?

Unger: Two of my uncles were good teachers who exerted a strong influence on me from the time when I was small. One was Count Karl von Arco, a representative of one of the oldest families in Austria. He taught me how to think in all instances and how to act when in difficulty. The other was Professor Gernot Eder, who taught me to think scientifically, analytically, metaphysically, and inclusively. Later, I enjoyed the guidance of a wonderful cardiac surgeon who was a fatherly mentor to me.

Ikeda: Those must have been splendid relationships. I can see that your youth was rich with intellectual development. Your father was a famous painter and the president of the Vienna School of Applied Art. What are your most vivid recollections of him? And which of the lessons you learned from your mother remain most firmly fixed in your heart?

Unger: I appreciate your asking about my late parents. My father was a painter with a very keen eye. He could accurately capture a landscape and render it in splendid colours – as you do in your photographs.

Ikeda: Your parents lived through both world wars and through the turbulent times of Hitler's *Anschluss*, the merging of Germany and Austria.

Unger: Yes, they were still very young at the time of the *Anschluss*. Father had to go to war, but they survived the fires of the conflict.

My father spoke clearly, extremely accurately, and honestly. From him I learned the importance of speaking correctly and justly in troubled times like the present. I imagine he learned this from his own father, who was an architectural engineer. My family taught me that money is not everything and that living justly is important. My mother was a cheerful, broadminded, optimistic person who taught me never to be afraid. No matter what problems she encountered, she would always say, 'It's nothing! There's bound to be a solution!'

Ikeda: Your parents were wonderful people, fine philosophers who taught you the path of happiness. I sense the great light of faith in them. True education consists in cultivating minds willing to struggle for the sake of justice.

Founded in 1990, the European Academy of Sciences and Arts has grown into an organization of 1,200 intellectuals from more than fifty countries in Europe, North and South America, the Middle East, and Asia. King Juan Carlos of Spain is a protector; Václav Havel, former president of the Czech Republic, is an honorary senator; and Mikhail Gorbachev, former president of the Soviet Union, is also an honorary member. I have had the honour of meeting all these people. Traditional academies exist all over Europe. Why did you decide to establish a new kind of academy?

Unger: The European Academy of Sciences and Arts was founded in response to the contemporary nexus of problems. We founders realized that the image of science has shifted away from the human being and has become lopsidedly material. We felt that, as materialism gets the upper hand in science and society, humanity diverges from its essential qualities and degenerates metaphysically, thus threatening to spoil human nature. Concern for this situation motivated the founding of the academy.

The natural sciences are rooted originally in humanity. The tendency toward materialism denies their spiritual aspects. Something similar occurs in the field of economics. Instead of serving human needs – including spiritual needs – science becomes disconnected from and domineering of humanity. Our academy approaches science from a different perspective.

Human relations consist in a harmonious, manageable triangle made up of relations with nature, relations among human beings, and relations with the spiritual. These three orient human existence and this orientation facilitates our ability to solve life's problems. All three relationships are essential, and none must be overstressed. Sciences that take nature as their theme – natural sciences and technological disciplines – require social and spiritual supplementation. The same is true for sciences that investigate relations among human beings – from social sciences and linguistics to law, economics and politics. Psychology and medicine build a bridge between natural sciences and social sciences and connect them with the purely spiritual fields of philosophy, art and religion.

Ikeda: In other words, human beings must occupy the centre of your harmonious triangle of natural science, social science and the humanities. In different terms, human beings do not exist for the sake of science; science exists for the sake of human beings. You said in a speech you delivered at Soka University, 'We must enable the sciences to work together to form a network encompassing all of life. . . . It is imperative for science to acquire a transdisciplinary philosophy.' What concrete actions is your academy taking to deal with this issue?

Unger: The academy is planning various projects for the present and the next few years in connection with problems of a sustainable society and investigations of water resources.

One of our big projects is the Institute of Medicine. We think that, in Europe, increasingly brutal capitalism, ridiculous public authorities, the bureaucratization of hospitals, and inept politics have created a two-class system of medical care. People with money can afford medical care; those without must look out for themselves. As a doctor, I find this set of circumstances intolerable. We have founded the Institute of Medicine in the hope of working together with the European Parliament and the Organization for Economic Cooperation and Development to develop a new healthcare system. We must not be satisfied with pragmatic – not to say, superficial – correctives but must look deeper into the problem. It is clear to all my colleagues at the academy that contemporary medicine has miscalculated the nature of its own existence.

Ikeda: In what connection?

Unger: Its strong inclination towards the natural sciences has caused medicine unintentionally to regard patients as cases that can be summed up in natural-scientific terms. The patient becomes an anomaly to be dealt with like other social anomalies; for instance, criminality. This invites the involvement of the authorities, since the patient as anomaly is a cost factor and cost factors must be minimized. Thus the patient becomes a mere object of medical and state action, and medicine acquires a meaning contrary to its true essence, which is to serve the interests of the patient and his or her good health exclusively.

Ikeda: Yes, you are saying that doctors are losing sight of the fundamental point that medicine exists for the sake of the patient. They subdivide fields of research and adopt the reductionist viewpoint that accumulated research in limited areas will provide them with a holistic view. This phenomenon observable in Western medicine can be seen in all the natural sciences. Studying only specific parts, losing sight of the patient as a whole, and considering patients not as human beings but as cases – anomalies, as you put it – may in a sense be inevitable since modern Western medicine developed within the framework of the natural sciences. But such an approach tends to neglect the living human being.

In a dialogue we conducted, the British historian Arnold J. Toynbee (1889–1975) said that 'it seems hardly possible for anyone to be a spiritually and ethically adequate physician unless he has some religious or philosophical view of, and attitude towards, human life and towards the universe in which mankind finds itself.'[2] Reforming medicine cannot be accomplished only within the medical field. It requires the help of religion, philosophy, ethics, politics, economics, and sociology.

Medicine directly concerns the totality of human life. That is why, first of all, we must make sure it includes the triangle your academy advocates: natural sciences, social sciences, and the humanities.

Dialogue among Philosophy, Religion and the Natural Sciences

Unger: Another related activity of our academy is the attempt to find common fundamental elements among fields of learning. Our philosophers and natural scientists willingly affirm this undertaking. Finding common denominators is the noblest task of science. As the science that implies the most radical contemporary social, ethical, and anthropological consequences, genetic technology provides especially powerful suggestions. As a doctor, I am occupied with the problem of when life begins and ends. In this issue, genetic technology plays a significant part.

Ikeda: Essentially Buddhism is an investigation of the fundamental issues of life and death and a search for ways to deal with the suffering arising from them. Both Shakyamuni and Jesus compared their roles as saviours with that of a physician. The Buddhist scriptures speak of Shakyamuni as the Great Healer; that is, a physician to life itself. Since illness is a manifestation of life activities, the wisdom of religion and medical science must be mutually beneficial. Dialogue between medical science and religion is indispensable to addressing the issue of genetic engineering and life ethics. Later I would like to refer to your great experience as a cardiac surgeon in this connection.

Unger: The need for dialogue is not limited to academic disciplines. Today, as information increases and is transmitted increasingly rapidly, dialogue becomes more essential than ever before. Whether they like it or not – whether they know it or not – cultures are interrelated. And this makes possible more intense exchanges. The exciting thing about intercontinental discussions is the way they enable us to compare our own traditional ways of thinking with those of other peoples. Serious comparisons of this kind always reveal more commonalities than differences. I could go as far as to say that we can reaffirm that we are the same in many ways.

Ikeda: I agree entirely. Through dialogue, people brought up in different philosophical traditions should work to discover more shared elements than differences. To date I have engaged in over 1,600 dialogues with leaders and thinkers of different religious and philosophical backgrounds including Christianity, Islam and Hinduism. One conviction I have gained as a result is that it is possible to form meaningful bonds of friendship with people of all spiritual backgrounds. Herein lies my unshakable confidence in universal humanism.

This applies to interfaith dialogue. With their long histories, world religions share many fundamental elements. In April 2002, I met with Abdurrahman Wahid, former president of Indonesia and leader of the largest Indonesian Islamic organization. We agreed that all religions exist for the sake of human happiness and that, without compromising their doctrines, they should cooperate in the name of peace.

Unger: Yes. Though they differ on some points, all religions share the same hope of peace for humanity.

Ikeda: At the United Nations Millennium Summit in the autumn of 2000, President Wahid said that dialogue can provide a human face regardless of ethnicity, cultural differences or historical backgrounds and pave the way for the promotion of common values and a commitment for a global culture of peace and harmony.[3]

Dialogue is the best way to help us discover our common humanity and return to universal human values.

Unger: As a doctor, I realize that human beings throughout the world are amazingly similar physiologically. On this physiology we impose violent cultural and religious differences. But given our bodily similarities, are we not fooling ourselves when we define cultural and religious differences as basic conflicts?

Ikeda: Physiological identity is one fundamental human equality that transcends race and ethnic group. Genetic technology has recently shown that individual differences in human DNA disposition amount to no more than 0.1 per cent. What is more, as Dr M.S. Swaminathan, the world-renowned Indian agronomist and former president of the Pugwash Conferences, insists, human beings and plants closely resemble each other genetically. In short, he concludes that modern genetics confirms the oneness of all life on Earth.

Unger: Scientific knowledge suggests that we should seek common elements and resemblances on different continents and in different cultures and religions.

Interfaith Dialogues as Spiritual Struggle

Unger: Dialogue serves to widen our horizons by calling attention to the experiences of others. But this is more difficult to do in the case of religion than in the case of natural science. In the natural sciences, measurements can be compared to provide reliable bases. A kilogram of bread is always a kilogram of bread. But spiritual bread cannot be so easily compared. Although no less important than precisely weighable material bread, it cannot be readily measured and sold. As we put too much faith in material bread, bit by bit, the material gains the upper hand over the spiritual until, ultimately, we have a capitalism in which humanity has metaphysically degenerated. Thus a means has been converted into a goal in itself. I object to this because capital must not be taken as a gauge for measuring the value of life.

Ikeda: The European Academy of Sciences and Arts has conducted a course on Buddhist–Christian dialogue as a way of countering materialism and has published reports on them in German and in English. It has also conducted dialogues among representatives of the four great religions Christianity, Islam, Judaism, and Buddhism. The significance of this great spiritual effort is sure to become historic. In response to your request, for three years starting in 1997, SGI participated in the six symposia held by the academy.

Unger: In spite of differences in systems of thought like those between Buddhist wisdom and Christian revelation, these six symposia were successful in ascertaining common points between the two religions that can be applied toward triumphing over problems confronting all humankind. We must build bridges.

Buddhism–Christianity Commonalities

Ikeda: Christianity and Buddhism share many things in common. First, both are salvationist religions. Buddhist compassion and Christian love both seek the salvation of humanity. The mission of Buddhism is the salvation of humankind through compassion that wells up from the cosmic life force and is made manifest in human beings. In the face of oppression, Shakyamuni and his disciples travelled the breadth of India uplifting the ordinary people.

Unger: Christianity teaches the love of God and strives to inculcate neighbourly love. Jesus Christ and his disciples suffered oppression and hardship as they taught the people. Jesus said, 'But I tell you, love your enemies and pray for those who persecute you' (Matthew 5:44). Thus, it teaches love for all humankind.

Ikeda: A second commonality is that both Christianity and Buddhism indicate ways of shedding light on human suffering and leading people to true happiness. Buddhism identifies fundamental, innate ignorance of the true nature of life as the basic cause of suffering. Christianity sees original sin as the source of all unhappiness. Both posit an eternal dimension to life where the causes of suffering are exposed and true happiness is attained.

Unger: As you point out, though their philosophical grounds for explaining it differ, both Christianity and Buddhism focus on the way through life: birth, ageing, illness, and death.

Ikeda: Their third shared characteristic is their teaching of the dignity of humanity and of life on the basis of the sacred. Buddhism teaches that the transcendent cosmic life force is inherent in each individual life as the Buddha nature and the dignity of life and of humanity derives from its manifestation. In the world civilization of the future, this dignity must become the basis of ethics and value criteria.

Unger: In the Christian view, human dignity derives from our having been created in God's image. In spite of their differences, Buddhism and Christianity converge on certain points transcending current problems; for instance, human rights, value systems, and the global environment.

Ikeda: When we met in July 2001, you proposed expanding the Buddhist–Christian dialogues to include all four of the major religions: those two plus Judaism and Islam. I was in complete agreement. SGI representatives took part in the first of the expanded conferences scheduled for 15 September 2001, shortly after the unexpected terrorist attacks on the United States.

Unger: The attacks were a great shock. Originally the conference theme was to have been life ethics. But at the beginning of the meeting, I said that the attacks perpetrated four days earlier increased the danger of global conflict. Against the background of this emergency, I proposed that the representatives of the four religions consider the topic of innate human destructiveness and aggressiveness. They did so. And, after lively debate, they reached a consensus to the effect that violent military retribution must be avoided and that discussion is the way to halt war.

Ikeda: I am grateful for the wise leadership you demonstrated at that time. I understand that the theologian and journalist Dr Norbert Göttler expressed the opinion that Buddhism can facilitate dialogues between Christianity and Islam. Repeated in succeeding years, these conferences among the four great religions have produced significant positive results.

Cooperating for Salvation

Unger: In Europe, very strong, even critical, secularization marginalizes religion with the result that European churches no longer have the powerful voice they once had. I by no means regard secularization as a misfortune. I see opportunities in it. Because it compels churches to justify themselves, it scuttles a great deal of the dogmatic ballast of the past. Only by showing that they are conducive to harmonious human coexistence can churches legitimize themselves.

Ikeda: You describe what we must require of religion in the twenty-first century. The lifeline of religion is empathy and cooperation in overcoming suffering. In a Buddhist scripture, concerning the cause of his sickness, the layman Vimalakirti says, 'Because all living beings are sick, therefore I am sick'.[4] Nichiren, too, declared 'The varied sufferings that all living beings undergo – all these are Nichiren's own sufferings.'[5] The spirit of accepting others' sufferings as one's own is surely the driving power that can propel interfaith dialogues, overcome differences, and build a symbiotic society.

Aurelio Peccei, a co-founder of the Club of Rome, and I discussed the possibility of global religious cooperation. He posed the pressing question: 'Have all these great religions ever called on the other major faiths to work and learn together how to guide the world population away from its present plight and towards terrestrial salvation before it is too late?'[6]

Unger: What are your thoughts on the main goals for interfaith dialogue today?

Ikeda: As a confirmation of the content of our discussion to this point, I might cite the following: First, religions should promote mutual understanding among themselves. This is the first stage of tolerance. If, while maintaining their distinctive systems of faith and philosophy, religions try to understand each other, they will be able to discover many things they share in common.

Unger: We have already seen how Christianity and Buddhism share common elements.

Ikeda: Yes. The second goal of interfaith dialogue should be for religions to learn from others and use these lessons for further self-development. All religions must grow in response to other philosophies and religions, the spirit of the times, and global conditions.

The third goal is, as Peccei pointed out, for all religions 'to work and learn together how to guide the world population away from its present plight'. Our current plight can be described in terms of three classes of relations – these being, as you point out, relations between human beings and nature, among human beings, and between the human and the spiritual. The first set of problems pertains to relations between humanity and the natural environment in the sphere of ecology: destruction of the ecological balance most dramatically represented by global warming. Then come problems of human-to-human relations in the political and economic spheres, including nuclear arms, conflict and war, economic disparity and poverty, and human rights and closely related terrorism. Central to these, the third category involves human spiritual decline and breakdown made apparent through phenomena such as apathy, outbreaks of violence, chronic mental illness, and drug addiction, even among the young. In this connection, ethical controls on rapidly developing information science and genetics are being discussed.

Challenging problems on a humanity-wide, global scale demands the application of results from many fields of learning. The key to everything is revival of the wisdom and creativity to enable us to make full use of our intellectual heritage. In all these efforts, religion has the mission of empowering the human spirit and elevating our ethical and moral values.

Unger: You have outlined in detail the way in which religion, in your precise understanding of it, can assist in opening the way to a peaceful world in which humanity and nature are recognized equitably. There can be no doubt that religion has the power to accomplish this.

Ikeda: The civilization of the twenty-first century must be a civilization of dialogue, as my friend, Harvard Professor Tu Weiming, has proposed. Professor Tu is active in many areas. For instance, in connection with the United Nations 'Year of Dialogue among Civilizations' in 2001, he spoke on Confucian civilization at a meeting of the Panel of Eminent Persons convened by Secretary General Kofi Annan. He has called dialogue an important mechanism for eliminating intercultural contradictions and collisions, and insists that we must recognize and respect others' values and conditions as well as learn and benefit from each other. By taking the lead in interfaith dialogue you act in accord with his admonitions.

Unger: Interfaith dialogue will become the cornerstone of a whole civilization of dialogue. Our own discussion shows that dialogue and mutual cooperation among all religions is both possible and desired by all humanity. Dialogue between Christianity and Buddhism and among the four great religions has inspired my determination to join you in continuing to speak out for the dignity of life, respect for humanity, and peace.

CHAPTER ONE

Religion and Tolerance

————————————

Ikeda: The virtue of tolerance is a vital issue of concern in this time of globalization and new encounters. Rapid improvements in transport and communications have significantly increased the possibility and frequency of encounters between people, to an extent that was unimaginable in the past. Getting to know our neighbours in the global village, we experience and are moved by diverse cultures and customs. Though very great differences are sometimes hard to bear, the world is coming closer and closer together with the result that we have no choice but to get along well with our new neighbours. In these circumstances, tolerance is essential. But the very word *tolerance* is rich with complexities and connotations. You have worked hard to cultivate a tolerant spirit. How do you interpret tolerance?

Unger: Tolerance is very active and very personal. It arises from discussions with others and presupposes a civilized viewpoint. It is the process in which I go out of myself to speak with others. It is an example of human arithmetic in which one and one make, not two, but three. When one opinion comes into real exchange with another, a third new opinion emerges, providing a starting point for further discussions.

Naturally, the spirit of tolerance faces limits when interpersonal relationships are strained. But these limits must be interrogated.

In any case, tolerance is a spiritual service to an other. It is an active form of coexistence in which I identify with and feel responsible for my fellows. Encounters with others lead to the greatest discoveries and maximum contributions to humanity. In a more than symbolic sense, each of us owes his or her existence to loving encounters between two people. Here is another example of human arithmetic in which one and one equal three.

Ikeda: Encounters with others broaden the individual and enrich cultures and civilizations. Meeting a stranger is in effect meeting an unknown self, with improving and revolutionizing effects on both parties.

But as history from very early times reveals, encounters can be characterized by tolerance and acceptance or intolerance and rejection. Intolerant rejections result in conflict and destruction, whereas tolerance leads to positive effects such as harmony and creation. In the terms of your human arithmetic, one and one can equal zero or even negative numbers, or it can equal three, four, five, and more.

The present age of unprecedented opportunities for encounters can produce equally unprecedented negative or positive results. It is deeply concerning that intolerant collisions are igniting the flames of war in many parts of the world today. The pressing task is to make the world a place of tolerance.

Capitalism, a New Religion

Unger: Today I perceive the existence of a new religion, the religion of capital. Though capitalism does not always entail the application of force, its language conceals its forcefulness with euphemisms such as letting people go, rather than firing them; restructuring, rather than closure; and flexibility, rather than irresponsibility. In spite of these circumlocutions, the language of capital has little to do with tolerance. It hardens hearts and holds humanity in contempt. Naturally capital is essential, but only insofar as it supports life. As the myth of King Midas reveals, capital provides no nourishment. Everything the king touched turned to gold, but he died because he could not eat the precious metal.

Ikeda: From the viewpoint of the Middle Ages, society today would seem to be surprisingly materialistic in its worship of capital as a god. Unquestioning belief in the omnipotence of the market is founded on the law of the jungle. The rich are superior to the poor, who are seen as inferior and must retreat from the social arena. The extent to which this makes people callous is concerning.

Globalization compels all humanity to live together in a shrinking world rife with inequality and injustice. People cast out by the indifference of others inevitably become deeply resentful. Terrorist acts are of course impermissible, absolute evils. But force of arms is not enough to eliminate them. Calling a halt to terrorism requires us to create a society characterized by empathy for the pain and sufferings of others. We must teach our society to sympathize.

Unger: Naturally we agree that aggression costing so many human lives is never justifiable. Self-critically, not in self-justification, we must ask ourselves how we can reduce the horrific hatred directed at capitalistic culture. The hatred has its roots in the way in which our economic system little by little subjugates the whole world and in doing so undermines the moral values of other cultures. This is why the 9/11 attacks were directed at a central symbol of capitalism. In this case, the hatred was combined with intelligence, rational execution, and endurance. The West now knows how it feels to be targeted and attacked. In the light of the dramatic post-9/11 developments, the word *tolerance* is pitiably mild and helpless. Nonetheless, we must not become resigned to fate. To do so would be to abandon belief in spiritual values.

Ikeda: I agree completely. The modern unmitigated market society is not the only system capable of surviving. Other societies, rooted in different traditions, have the potential of diverse forms of development. To uncover these inherent potentials we must promote the spirit of tolerance to every corner of the globe. In Buddhist terms, the spirit behind the spreading of such tolerance is compassion.

Charters of Tolerance

Ikeda: Tolerance is indeed a fundamental challenge of the twenty-first century. In the spirit of interreligious dialogue, in January 2002, the European Academy of Sciences and Arts issued its Tolerance Charter, which I should now like to discuss. It proclaims tolerance to be the duty of all people. How did the document evolve?

Unger: Our Tolerance Charter can be understood as having evolved from the multi-disciplinary and international nature of the academy. Our many religious dialogues have shown coexistence to be both a virtue and a pressing demand. The desire to realize this necessity led to the creation of a council to formulate the charter as representing our senate and speaking for the academy. For us, it has great cultural significance.

The Charter comprises two sections: the preamble and six definitions. The preamble consists of three paragraphs, which respond to the current state of society.

Ikeda: The preamble, as I understand it, states that everyone must respect the differences and values of others as valuable in themselves.

Unger: Our charter describes human life as constantly changing and sees the speed and scale of change as characteristic of our time – so much so that the desired direction is obscure. It adds the injunction to employ a tolerant culture to counter diverse forms of intolerance. Growing individualization leads many people to rate their own lives as absolute and to be unsympathetic with others. The family ought to function as the stable nucleus of human society. But, imperilled, it is increasingly incapable of doing so. We therefore remind all official personages of their responsibility to make tolerance valid as a social good. As such, we have called out to all people, regardless of their religious or political denominations.

Am I correct in understanding the preamble of your SGI charter to express Buddhist humanism?

Ikeda: Yes, the SGI Charter, enacted in November 1995, sets forth our fundamental ideals. Its preamble states in part:

> We, the constituent organizations and members of SGI . . . being determined to raise high the banner of world citizen-ship, the spirit of tolerance, and respect for human rights based on the humanistic spirit of Buddhism, and to challenge the global issues that face humankind through dialogue and practical efforts based on a steadfast commitment to non-violence, hereby adopt this charter . . .

The spirit of tolerance is the guidepost of the charter. It is followed by ten articles stipulating SGI's purposes and principles. Buddhism aims to develop a universal humanity, thus the charter rests firmly on this goal. Our membership comes from 190 countries and territories and transcends national, ethnic and cultural boundaries. The diverse ethnicities, cultures and religions in the regions where we are active have potentials for both tolerance and intolerance.

Through the stimulus of open dialogue and an abiding contri-bution to the local community we strive to encourage benevolent tolerance. We believe that the culture of dialogue provides the soil in which universal humanity can bloom in diverse forms. The SGI

Charter expresses our vow as Buddhists to encourage peace and symbiosis. It is in this spirit that I have engaged in dialogue with leaders and thinkers from many cultural and religious backgrounds, including Christian, Muslim and Hindu.

Unger: I understand. And that is why I am eager to join you in the pursuit of the spirit of tolerance for the sake of peace.

The preamble of our Tolerance Charter is followed by six definitions:

1. Tolerance is the individual readiness to stand up for the dignity of any other human being.
2. Tolerance is part of a value system that emphasizes human dignity.
3. Tolerance requires a person's ability to understand other people and to respect different behaviour.
4. Tolerance is based on a self-confident point of view.
5. Tolerance serves as a protector of dignity and freedom of each human being within his or her own cultural environment.
6. The guaranty and continuing development of tolerance should be considered our common duty and the fundamental element of any education.[1]

Ikeda: I agree totally with the spirit of tolerance set forth in these definitions. The following articles demonstrate the way the spirit of tolerance finds clear expression in the main text of the SGI Charter:

- SGI, based on the ideal of world citizenship, shall safeguard fundamental human rights and not discriminate against any individual on any grounds.
- SGI shall, based on the Buddhist spirit of tolerance, respect other religions, engage in dialogue and work together with them toward the resolution of fundamental issues concerning humanity.

- SGI shall respect cultural diversity and promote cultural exchange, thereby creating an international society of mutual understanding and harmony.
- SGI shall promote, based on the Buddhist ideal of symbiosis, the protection of nature and the environment.
- SGI shall contribute to the promotion of education, in pursuit of truth as well as the development of scholarship, to enable all people to cultivate their individual character and enjoy fulfilling and happy lives.

As these statements make clear, the academy's Tolerance Charter and the SGI Charter generally coincide in their approach to the spirit of tolerance.

Unger: Our approach to dialogue is one of the many things we share. Any sincere dialogue indicates the kind of tolerance represented in the New Testament appeal to go the 'second mile'. This difficult metaphor refers to the Roman requirement made on all Jews to accompany a Roman citizen for a mile to carry his baggage. Christ's injunction to go a second mile lays the groundwork for dialogue, since sharing the road for the additional distance gives both parties a chance to talk things over.

You come from the Buddhist tradition, and your image of the world differs from the Christian one. Nevertheless, we Christians are becoming increasingly familiar with that image. Not that we want to accept it entirely, but we discern its profoundly humanistic elements. The traditions behind your charter and our Tolerance Charter differ. We do not associate tolerance with our religious tradition as you do; we are more intensely secular. Still we are connected by basic values, and working together reveals new elements.

Ikeda: The content of the academy's Tolerance Charter should serve as a model for the leaders of all nations. Overcoming intolerance is one of the primary aims of the United Nations, as is stated in the preamble to its charter: 'to practice tolerance and live together in peace with one another as good neighbors'. In accordance with this, the UN held the World Conference against Racism, Racial Discrimination, Xenophobia and Related Intolerance in Durban, South Africa. From the preparations stage, SGI representatives participated enthusiastically in all phases of the conference. In spite of international efforts of this kind, however, the supression of human rights through intolerance and discrimination is proving difficult to uproot.

Unger: Undeniably, things like racism stand in the way of tolerance. You and I both adopt human rights as our starting point. We shall have much more to say on the topic later in our dialogue.

Monotheism and Intolerance

Unger: In thinking of intolerance in the world today, I am horribly saddened by the inhuman contention in the Middle East, the very cradle of the three great Abrahamic religions: Judaism, Christianity, and Islam. All three profess an omnipotent God recognized by the patriarch Abraham but have in reality entombed his spirit. For me, the facile silencing of insights that might question one's own ideas and endeavours encapsulates intolerance. We must always be on the lookout for intolerance in the garb of hypocrisy. Power plays can easily be tricked out in humane phrases.

Ikeda: Although history shows that monotheism is more prone to intolerance than polytheism, the tendency is not always manifested. Polytheists too can be intolerant. It is not only teachings but also the attitudes of their believers regarding those teachings that determine intolerance or tolerance. Indigenous Japanese Shinto is polytheistic. When it became the state religion allied with fascist ideology in the twentieth century, however, it manifested the narrowest intolerance, thus bringing down calamity on the peoples of Asia.

During World War II, Tsunesaburo Makiguchi, first president of Soka Gakkai, and Josei Toda, second president, refused to worship Shinto gods and were imprisoned as a result. Makiguchi died in prison; Toda was released after two years. Though apparently more tolerant, depending on their application and the attitudes of their adherents, polytheisms too can demonstrate extreme intolerance and violence.

While it is true that Buddhism has never experienced blood wars of religion and has generally been tolerant, many Buddhists approved Japanese ultra-nationalism and readily made compromises and cooperated with fascist ideology. Lack of a strong pacifist spirit prevented such Buddhists from resistance to the regime. Only a few like Makiguchi and Toda stood up courageously against the authorities, insisting on freedom of faith even when imprisoned for their stance. The person defines the principle.

Examined closely, wars described as religious conflicts very often arise less from causes inherent in the religions themselves than from conflicts of political and economic interests. Today we see complex historical conflicts and profit–loss antagonisms simplified to dualistic formulas of religious good and bad. It can be said to be a form of abuse.

Unger: We must always discern whether religious opinions are in fact the cause of a given conflict. Whereas the warring parties may have specific religious backgrounds, this does not justify blaming religion for the fighting.

Ikeda: Very true. Moreover, one believer's resorting to extreme violence provides no grounds for suspicion and hostility toward all fellow believers.

Unger: Since religion is a powerful factor in life, we ought to prevent its leading to intolerance. But history casts doubt on our ability to do this. For instance, while teaching neighbourly love, Christianity is sometimes intolerant, as is illustrated by things like the Crusades, the persecution of Jews and the abortion issue.

Western history reveals irreparable errors. In justifying the Crusades, Pope Urban III branded Muslims as demons. The Crusaders caused Muslims untold suffering, thus giving cause for hostilities between the West and the Middle East and making mutual understanding difficult. The demon label stuck. Once one party begins regarding the other with animosity, the scene is set for multicultural social conflict.

Ikeda: Demonization is a first step to war. When other people are considered not as equals but as non-humans insusceptible to suffering and pain, it becomes possible to attack and harm them unhesitatingly. They are catagorized into abstract entities devoid of individual expression. When such an attitude becomes firmly established, it is easy to draw up and expand schemes of antagonism. Theories of the clashes of civilizations, as with hostile attitudes toward Arab or Chinese civilization, entail this kind of risk.

The world owes Islamic culture a great deal. By importing what was at the time the most advanced natural-scientific thought and other disciplines from the Islamic world, Christian Europe was able to enjoy the so-called twelfth-century Renaissance from which modern Europe was born. The world must be more keenly aware of the great benefits the Islamic civilization has brought.

Unger: What are your opinions on the idea that, by its very nature, religion engenders tolerance?

Ikeda: You bring up an important point. Essentially, I do not believe that religion itself threatens tolerance. The crisis of tolerance is not intrinsically religious but is a complicated intertwining of political and economic confrontations. Religion must battle against this crisis and strive for a popular consensus of tolerance.

Discussing the issue of religion and tolerance and intolerance requires an investigation into the nature and goals of religion itself. Religion should guide people toward eternal, universal truth. This being the case, all religions ultimately should incorporate the spirit of promoting human happiness. A religion denies its own religious nature when it becomes intolerant of others and self-righteous.

Unger: Throughout history, devotees of some religions have promoted intolerance in their desire for an undisputed claim to ultimate truth. The same thing happens today. Some people even believe that intolerance is primordial and strong, whereas the tolerance opposed to it is a later cultivation.

Ikeda: Religious devotion requires firm faith in the perceived truth of one's religion. But to allow this faith to make one intolerant, exclusivist, and narrow-minded runs counter to the nature of religion, which is to elucidate truth. Instead, people of religion should pool their wisdom for the sake of revealing truth and compete with each other in realizing real human happiness. Discovering in other religions truths perceived in one's own reinforces one's sense of correctness. To reject the truths inherent in other religions in their entirety, on the other hand, maybe tantamount to reducing one's own credibility.

Unger: In this connection, consideration of the European age of Enlightenment may prove useful. Enlightenment thinkers believed that the ultimate achievement of humanity was the use of reason to expose and reject unconditional insistence on absolutes. In Austria, Emperor Joseph II is intimately connected with this issue because of the Patent of Tolerance he promulgated promoting religious toleration.

Ikeda: Joseph II, who was Mozart's protector, was known as a highly progressive emperor for his radical reforms.

Unger: We should also remember that the Austro-Hungarian monarchy was the first political body in the nineteenth century to recognize Islam as a state-sanctioned religion.

Ikeda: Austria was a good example of the culturally enriching effects of a tolerant attitude. By welcoming the talents of peoples of all religions and races, Austria made the imperial capital Vienna the cradle of new learning and art. Among the Habsburg rulers were several great art lovers, like the emperor Maximilian I. It is even said that a Habsburg family precept was that the harp is mightier than the sword.

A deep cultural understanding stimulates the creation of an atmosphere of tolerance. In addition, from ancient times, the human spirit has created cultures that take their sources in religion. This is why tolerance of other cultures relates to tolerance of other religions and of other people.

Though I digress, I would like to mention the visit of the Vienna State Opera to Japan in 1980 at the invitation of Min-On Concert Association. It was an unforgettable event for me as the founder of the association.

Unger: It was a truly historical event. In Europe, tolerance is often viewed as a philosophical rather than a religious ideal. Is that true in Japan as well?

Ikeda: In the West, tolerance is understood to be a philosophical ideal. Traditionally, in English and other Western languages, words standing for ideals and ideas represent philosophical concepts. In recent times, with increasing secularization, traditional theological answers to existential questions have lost their power to convince, giving rise to existential philosophy since the time of Søren Kierkegaard (1813–55).

The Japanese, on the other hand, lack philosophies like those of the West that pursue eternal, universal ideals, so the attempt to determine whether the issue of tolerance is religious or philosophical itself poses difficulties. Since in the East religion is a philosophy of humanity, life, and actual practice, a philosophical issue is naturally viewed as a religious issue. Insofar as tolerance is essential to harmonious coexistence, it must be guided by religion.

I find such guidance in the Buddhist philosophy of the dignity of life, the equality of all people, and respect for humanity. In this connection, the example of the bodhisattva named Never Disparaging, who appears in the Lotus Sutra, is especially instructive. Never Disparaging paid reverence to the supreme value – that is, the Buddha nature – inherent in everyone. Ironically, unaware of this Buddha nature within themselves, the very people he revered despised and persecuted him. Nonetheless, he persevered in revering them until ultimately they were enlightened to their own supreme dignity.

Unger: Religion's greatest task is to restore to people a sense of their own worth, be it through prayer, meditation, fasting or introspection. The re-realization of self-worth through religious practice can be the foundation of sustained harmonious coexistence.

Ikeda: Yes. A person profoundly aware of the dignity of his or her own life respects the lives of others and works with them to manifest individuality and pioneer a life of value.

Unger: It is counter to the nature of religion to incite conflict. Essentially religions should show humanity the way to freedom and avoidance of evil for the sake of a just coexistence. Religions concern themselves with reining in human evil and, as in the case of

Christianity, preparing people for salvation. Salvation-seekers must never resort to force.

Ikeda: As you indicate, world religions that have served as foundations for great civilizations all essentially embrace ideals of peace and coexistence and have great potential to realize these ideals. Dr Majid Tehranian, who is Iranian by birth, is director of the Toda Institute for Global Peace and Policy Research, which I founded. He suggests this great potential in his comments on Islamic *jihad*. He makes the point that the word *jihad* has at times been interpreted as 'holy war' in Islam, but in the strictest sense, the *jihad* that relies on armed force is external and minor. The internal *jihad* is understood to be the process of spiritual purification through which the evil that resides within, such as greed and hatred, is overcome. It is the latter that is deemed to be the Great *Jihad*. Another point he raises is that Islam teaches that the use of force may be permitted only in self-defence.

Unger: Tolerance is grounded in the Abrahamic-Christian traditions and is to a great extent obligatory in the religions in these traditions. Interestingly, the three keywords of the French Revolution – liberty, equality, and fraternity – are aspects of tolerance. This caused the church – especially the Roman Catholic Church – difficulties as it was compelled to address the revolution's ideals. Tolerance grounded in the Christian tradition and liberty are obviously closely related to each other.

Ikeda: Although the French Revolution still has not been definitively evaluated, it was undeniably a major step toward human liberty. In your view, then, this development can be seen, from the standpoint of the Roman Catholic Church, as a manifestation of the God-given virtue of tolerance. We must continue to promote open-minded dialogue among religions to evoke and strengthen this virtue.

Religion and Authority

Unger: The professional clergy can be a stumbling block to the spread of tolerance. This is because they fear the undermining of their authority and so, like everyone who is afraid, they resort to threats. They menace with damnation anyone who does not adhere to their views.

Ikeda: An authoritarian clergy is degenerate and has lost sight of salvation as the basic mission of religion. All the great founders have established their religions in response to the cries of suffering contemporaries. The teachings and clarifications of life and of the universal truths they have evolved from their own ideas have spread and resonate with the hearts of the people because they address popular needs.

Nonetheless, once rigid social systems and positions have been established, some clergymen begin using the people as mere means to protect their own advantages and prestige. When this happens, they are prone to ignore the popular voice. If the great goal of salvation for the masses falls from sight, all desire to improve and all wisdom to see the truth are lost. This lowers levels of religiosity and results in false, tyrannical priests more secular than the secular. Unless we struggle resolutely against such false priests, the people will fall upon the road to misery.

In my youth, State Shinto, the psychological mainstay of the ultranationalists, hurled the people of Japan into a war of aggression. As you know, Soka Gakkai first and second presidents Makiguchi and Toda unrelentingly wrote and spoke out against the authorities of their day. Inevitably, after Japan's defeat, I too was critical and highly mistrustful of religion yet was in search of a correct philosophical attitude toward life. At that point I met Toda, my mentor in life, when I was nineteen years old, and was so strongly attracted by his profound humanity that I came to embrace faith in the Buddhism of Nichiren, to which he was devoted.

From our very first meeting, I felt as if I had known him for years. With great sincerity and candour, he urged me and my contemporaries to create a nation with a peaceful, harmonious culture, a country that could contribute to the wellbeing of all humanity. From Toda I learned that true religion serves the interests of peace-building, cultural development, and the happiness of the ordinary people. I understood fully that it generates indomitable faith, action, and humaneness.

Unger: In keeping with Toda's admonition, you now devote yourself tirelessly to promoting peace through cultural exchanges that are indispensable to the peaceful future of humanity.

Ikeda: I am all too aware of the perils authoritarian religion poses, not to mention the dreadful nature of government authority's exploitation of religion as a political vehicle.

In one of his works, the Austrian writer Stefan Zweig (1881–1942) described the theologian Sebastian Castellio (1515–63), who issued a declaration of tolerance, saying that 'persons in authority always endeavour to justify their deeds of violence by appealing to some religious or philosophical ideal'.[2]

Unger: Yes. That is an historical fact.

Ikeda: Zweig insisted that 'violence debases the thoughts it claims to defend'.[3] Religion is destroyed from within when it allows itself to be used by or actually cooperates with authoritarian violence and belligerence. Authority has an inherent demonic egoism. In contrast, true religion essentially acts to relieve human suffering. Whereas authority oppresses, dominates, and trivializes humanity from without, religion evokes the force of life from within, enabling human beings to manifest great individuality capable of creating great value. By its nature, religion is an art for purifying the egocentric greed of authority and orienting political power towards the interests of ordinary people.

Unger: Yes, politics should be an art of service to the people. After all, the word *minister* derives from a Latin term meaning 'servant'.

Ikeda: Religion should be a force for transforming authoritarianism. Sometimes, however, its desire to protect itself by means of political authority or to become an authority and a power in and of itself generates intolerant cruelty. Because, as you say, religion is an essential part of human nature, hypocrisy on the part of religious leaders invites profound distrust and confusion.

The tyranny of false religious leaders hiding behind authority is unacceptable. We must rely on such peaceful means as free speech to resist intolerance that threatens human dignity. This is the meaning of positive tolerance, the spirit of which is to encourage ordinary people to become stronger and wiser so that they won't be deceived by intolerant authorities. This is the kind of work we are engaged in.

Unger: Precisely. People with tolerant minds have the courage to advance the cause of peace. They can resist the fanaticism and vengefulness that threaten human dignity. It is a mistake to assume – as I said earlier – that intolerance is primeval and strong, whereas tolerance is derivative and weak. Tolerance and religion are inseparable.

Ikeda: I might cite a historical reference substantiating your assertion of the inseparable nature of religion and tolerance. Some of the leading intellectuals in the world I have talked with have demonstrated keen interest in the ancient Indian king Ashoka and his humanistic politics founded on Buddhist compassion and conducted in the name of the general happiness. An advocate of tolerance, he recognized others' freedom of belief. In addition, abandoning wars of conquest, he carried out extensive peaceful exchanges with many other countries. His actions demonstrate the ideal relation between politics and religion and provide a shining example of the way religious tolerance promotes a flourishing prosperity. Ashoka politically embodied the spirit of Shakyamuni, the compassionate ideals that were the origin of Buddhism.

In all religions, constantly referring to the spirit and basic teachings of founders is of the greatest importance. Ashoka opposed religious leaders who had abandoned the spirit of Shakyamuni and had forgotten their responsibility for saving ordinary people. One of the many edicts he had carved on stones and pillars as a way of communicating with the people is this: 'Whoever, whether monk or nun, splits the Sangha [the community of Buddhist believers] is to be made to wear white clothes and to reside somewhere other than in a monastery.'[4]

Toda was fond of telling us that, if they should assemble in one room, the founders of the world's great religions and philosophies would certainly find much to agree upon. All strivers for human liberation and peace – among them, Jesus, Muhammad, and Shakyamuni – supported the afflicted and helped the suffering, as did Nichiren, whom we revere.

One aspect of the movement founded by Nichiren was the admonition to return to Shakyamuni, and his teachings, for those who sought to free all people from suffering. It was also the admonition to return to the Lotus Sutra, which teaches the dignity of all human beings and awakens them to this truth. The struggle which Nichiren undertook then was to return to the starting point of religion, which lies in alleviating human beings from suffering and inspiring them to geniune happiness.

Harmony with Scientific Thought

Unger: As a scientist, I consider tolerance the basic condition for coexistence. In their own way, religions are science. They generate theologies to guide believers in the search for God and in this way provide new stimuli.

Ikeda: Religion is the wise pursuit of the truth of life and the universe. It is an existential search for ways to realize human happiness. On the basis of the results of their search, all religions and denominations have developed organized systems of teachings in the form of theologies and doctrinal studies. Using their own religious experiences, founders and believers evolve consistent and coordinated theoretical systems. In this sense, religions embody the empirical logic required of modern science.

Makiguchi was an educator. His desire for the happiness of all children inspired him to define the values that they ought to manifest in life and to build his own original theory around them. During the course of this undertaking, he encountered Nichiren Buddhism and was so sympathetic with its ideas of the dignity of humanity and the possibility of reforming real life and society that he became a believer. This was the origin of the Soka Gakkai's popular movement for peace, culture and education, which is founded on Nichiren Buddhism.

As a modern man, Makiguchi set out to prove the universal validity of Nichiren Buddhism through scientific methods involving actual practice and verification. His approach was a modernized version of the method Nichiren himself used to compare religions and philosophies: documentary proof, theoretical proof, and actual proof. As I have indicated, Buddhism, especially the Buddhism of Nichiren, is rational and empirical in the manner of modern science.

Unger: That's fascinating.

Ikeda: A religion that serves as a basis for twenty-first century civilization must both harmonize with and lead scientific rational thought.

I believe that Buddhism includes ideas that resonate with modern observations found in the social sciences as well as in humanities. My discussions with intellectuals from all over the world have confirmed me in this belief.

In any event, religions in the twenty-first century must address the question of how human beings should live in the most humane manner possible, based on clear perspectives on the world, society, and the natural environment.

Unger: For me, the fundamental question in life is, where we have come from and where we should be going. Religions, philosophies and the arts have all striven to find the answers to this question and have tried to do so in their own ways. Nevertheless, the same questions are being asked to this day and different answers are given. I find some of them unreliable, unconvincing, or outright wrong.

When it comes to the issue of tolerance, we need to be articulate. This is because I could not say I truly examined an answer if I do not make a value judgment on it. Judging an answer's value, however, is different from criticism. I think we need to be clear about this.

Ikeda: The more critical the issue, the more we require candid dialogue.

What are your opinions of the assertion that, from the seventeenth century, in Europe, with the reaction against religion and the rise of rationalism, the established religions had retreated to be replaced by scientific rationalism, which in its turn became a new kind of religion?

Unger: Certainly in the last 200 years, the mathematical and rational have been overvalued. The prevailing worldview is eccentric, unregulated and abnormally inclined in the direction of the rational and theoretical. Excessive stress on money is synonymous with the distorted worldview. Regulations have been relaxed in my own field of medicine, too. In medical schools, human beings are looked down on as mere objects. Patients are treated as prospective payers of medical insurance.

We must firmly realize that science is not and cannot be a replacement for God. Exclusive emphasis on reason has essentially altered relations between human beings and nature, among human beings, and between the human and the spiritual. We have fallen out of an anthropocentric harmony. Concerned people are beginning to realize that there are many things in heaven and earth that cannot be measured. Like all other worldviews and cultures, the new global culture must clearly rest on human beings' embrace of faith. Though subject to diverse interpretations, religion is the source of all culture. That is why inter-religious dialogues of the most substantial kind are essential.

Ikeda: From this viewpoint, interfaith dialogues must be made the core of intercivilizational dialogue.

Unger: Very true. In the extreme secularization of the present, owing to the limitless liberty of pluralism, as long as one's activities are not illicit, each person is completely free to act according to inclination. Undoubtedly this set of circumstances contributes greatly to the exploitation and destruction of the planet at the hands of human beings. These developments are moving ahead in tandem with what might be called politically correct tolerance as a mere formality, which is patently insufficient. In the defiled world of today, tolerance with internal content is called for.

Ikeda: Yes, I can see that. Formalistic tolerance that gives free play to the egoism of those on each side of an issue is impermissible. True tolerance is founded on overcoming rapacious human greed. This is a role religion has played since ancient times.

Tolerance that does no more than recognize the existence of others is basically passive. I define active tolerance as that which enables us to bring to fruit a richer humanity by respecting, delighting in, and learning from the existence of others. It is tolerance with – to use your words – internal content. Passive tolerance is mere formality. Whatever is called tolerance must be examined from the standpoint of whether practising it makes people more deeply compassionate and happier.

Unger: Sustaining the world requires human beings to come to terms with their own essential natures. Because this must begin with the individual, we must uphold three moral stances: human dignity, human development and human protection – the protection of life. Thoughtful consideration of the value of life is essential to the last of the three. Interfaith dialogues can work out basic policies for this and for the protection of humanity and society, creating global values transcending religious differences.

Ikeda: Indeed, interfaith dialogue is one of the pressing needs of our times. Tolerance with internal content spurs one to open-minded dialogue and mutual learning. On the other hand, a closed attitude of simply abiding a heterogeneous 'other' indicates formalistic tolerance.

Your three models represent three indispensable guidelines for transcending greed and cultivating tolerance.

Severing the Chain of Violence and Hatred

Unger: As we agree, realistically, we always have to deal with intolerance. It is possible to debate the innate nature of aggression and intolerance forever. In Europe it has always been thought that human beings are by nature good and society makes them aggressive. Purportedly peaceful primitive peoples have been used to substantiate this opinion. The majority of ethnologists now, however, are of a different opinion. Primitive peoples too can be aggressive. Tolerance must be cultivated like such other cultural skills as table manners and courtesy. It is all a matter of education.

There is, of course, an element of chance. Education can succeed only when parents set an example. Everybody knows how hard it is to lead an exemplary married life. Children who observe their parents trying to be virtuous will later try to emulate them.

Ikeda: We agree that education is what makes human beings truly human. The important issue is the nature of education: Does it result in a culture of violence or a culture of peace? Though they are deeply rooted in human life force, aggression and violence do not necessarily constitute the true human essence. Humans innately possess the disposition to suppress, control, and sublimate violence and aggression.

In its profound search into human life, Buddhism discovered both fundamental good in the form of the Buddha nature and fundamental evil in the form of ignorance, both constantly in conflict with each other. This observation indicates the need for human beings to be aware of the violence within them and constantly strive to control it. Similarly, we must always try to evoke tolerance to oppose the tendency toward intolerance. Any laxness in the ceaseless spiritual struggle can allow barbarous, intolerant violence to get the upper hand. This is true on the level of the individual and on that of whole civilizations as well.

Anyone who combats violence with violence sinks to the level of and psychologically unites with the violent opponent. This mistaken means cannot be justified even when applied toward the attainment of a righteous goal. A lofty goal requires correspondingly noble means. Violence threatens the dignity of life. It must be met with non-violence, which rigorously guards that dignity.

Unger: Tolerance cannot be measured. It contains something unpredictable. A person predictably reacts against another who causes him pain. As is seen in the Middle East today, this starts an escalating chain reaction of violent counter-reactions against counter-reactions. But the creativity of a tolerant reaction breaks this predictable circle.

Ikeda: A very important point. To sail always in the direction of lasting peace, the ship of human history must ride a powerful current: the unseen but mighty flow of transformation in the mind of humanity. The best-devised treaties and structures cannot be expected to produce desired results without a change in the minds of the people themselves.

Unger: My experience with interfaith dialogues indicates that charisma is more productive than mere words. Different people mean different things by words like *virtue*. The words themselves are less important. What counts is the actual conduct of a person of virtue. Tolerance cannot be systematized because it is too individual and intimate. It is always a personal issue.

Ikeda: Certainly the actual practice of tolerance is more important than talking about it. Further, to give substance to tolerance, a revolution of the underlying human spirit is indispensable. Discussions of freedom, equality, and tolerance tend to end up as conventional examinations of systems. Without a change in the underlying spiritual condition of society, any solutions formally produced may simply amount to a form of coercion. This would only breed further social tension and intolerance.

The twentieth century saw intolerance embodied in the totalitarian state, which led to tragic revolutions. Carl G. Jung stipulated that 'if only a worldwide consciousness could arise that all division and all fission are due to the splitting of opposites in the psyche, then we should know where to begin'.[5] With this general awareness, we can discover where to begin dealing with the issue. I believe that we must start with internal revolution.

Images of Freedom and Responsibility

Ikeda: The core Buddhist doctrine of dependent origination means that everything human and non-human is interdependent and nothing exists in isolation. This in turn means that nothing in the world is unrelated to the self. One's every action is closely connected with the dynamism of cosmic creation. In connection with tolerance, too, we must engage in a reformation of the inner life that leads to an awareness of interdependence and evokes a sense of responsibility for others.

Unger: Very true. Awareness of interdependence evokes a sense of responsibility. Perhaps in talking about tolerance we tend to over-emphasize freedom. At the end of the nineteenth century the Statue of Liberty was erected in New York to tell wanderers they had reached a place of freedom. I vividly remember a speech delivered in 1964 at the University of Vienna by the Austrian neurologist and psychologist Viktor Frankl (1905–97) who said that the importance of responsibility should receive greater attention and that it would be a good idea to erect a Statue of Responsibility on the West Coast of the United States.

Ikeda: The sense of responsibility for others and for society must be evoked internally. The English word *responsibility* means an ability to respond. People suffering intolerant suppression want more than freedom or preferential treatment. They want society to respect, value, and acknowledge them. They want solidarity as a member of society with the ability to make responses. Responsibility of this kind is the proof of true freedom. The only way to actualize tolerance is to increase the numbers of people with such a sense of responsibility and to make solidarity global.

Unger: In connection with responsibility, I was deeply impressed by the words Cardinal Franz König (1905–2004) spoke in the closing address of a meeting of the European Academy of Sciences and Arts and the Institute of Medicine [now called National Academy of Medicine] in Washington DC. He said that researchers must do their work honestly and that, whatever materials they use, scientists must follow the voice of conscience. I might put it this way: Does not responsibility equal the voice of conscience telling us where we ought to go?

Ikeda: Tolerance entails listening to our inner voice or conscience. It is dialogue with both other people and with the self in a ceaseless inquiry into the possibility of one's prejudice and self-interest. Viktor Frankl said that the voice of conscience is the voice of a transcendent being, of God. Having discussed tolerance, let us examine the religious dimension of love and compassion.

CHAPTER TWO

Buddhist Compassion and Christian Love

Ikeda: Buddhism can be described as a religion of wisdom and compassion. It is characterized by cosmic wisdom founded on Shakyamuni's enlightenment to the fundamental Law, or truth, of the universe and cosmic compassion combined with that wisdom. The Buddhist concept of compassion contains two aspects: the desire to share a friendship with others (Sanskrit *maitrī*) and the desire to embrace the suffering of others as one's own (*karunā*). Thus, it originally means supreme amity for people without discrimination of any kind. On the basis of these fundamental meanings, Mahayana Buddhism interpreted compassion as giving joy and relieving pain.

Nichiren considered these two aspects of compassion to represent the model-setting love of a father and the embracing love of a mother. By setting an example, a father encourages his children to do what is right toward others. Referring to the words of the Chinese Buddhist Zhang'an (561–632): 'If one befriends another person but lacks the mercy to correct him, one is in fact his enemy.'[1] Nichiren encouraged strictly remonstrating against the evil of others.

The embracing, motherly aspect of compassion, on the other hand, accepts everyone exactly as they are and sympathizes with their joys and sorrows. Compassion, then, is manifested in a practice that maintains balance between both aspects while seeking improvement for the self and the other.

What, in your view, are the meanings of Christian love and commiseration?

Unger: Our religious schools concentrate on mercy, a central ethical concept in Christianity, as in Buddhism. Mercy means putting oneself in the other's place, having commiseration, and offering help. In my opinion, mercy has little to do with tolerance. Mercy is on a different plane requiring a different virtue – that is, charity.

If they are different virtues, how do mercy and tolerance interrelate?

Imagine that one man has caused another pain. The sufferer stimulates my desire to be merciful; I want to help him. Also I want the inflictor of pain to leave his victim alone. It can be, however, that the wrongdoer himself has grounds for behaving as he does. Perhaps his victim has previously insulted him. In such a case, I must appeal to his tolerance. But the situation can work out well only when I appeal to the tolerance of the victim too. Otherwise there will be no end to the contention.

Mercy is related to the protection of tolerance, which can inspire willingness to forgive. It inspires me to action, which can only be crowned with success when both involved parties actively strive to resolve the issue between them. Both sides of the conflict must cope with the difficult task of forgiving. Success has its price.

The great value Christendom attributes to forgiveness is apparent in the Lord's Prayer: 'And forgive us our trespasses as we forgive those who trespass against us.'

Ikeda: You put mercy and tolerance on two different planes. Mercy is on the level of human sympathy, whereas tolerance is part of the loving virtue derived from God.

Unger: Exactly. My experiences with interfaith dialogues have convinced me that tolerance generates useful values. All actions should work to protect good and avoid evil. People ought not to be miserly, envious or arrogant. Charismatic virtues like faith, love, and hope describe this posture in a positive fashion. Thus I see tolerance mainly as a virtue of coexistence. On any scale of values, the main theme is warding off evil. The highest among the charismatic virtues is love.

Of course, wisdom is important for evaluating and considering matters. For instance, a judge who combines wisdom with great experience will be lenient in punishing minor offences, perhaps being satisfied with an admonition when another judge might call down the full weight of the law.

Whence comes wisdom? Can it be learned? Certainly, since it does not arise solely from life experience; it can be taught, but only by charismatic teachers. The best organized schools and institutions can neither guarantee nor compel charismatic wisdom. Wisdom is bound to the individual. Organizations can reward – or sometimes hinder – the work of wise teachers but remain always vessels that, though full of it, cannot create wisdom.

Ikeda: Certainly wisdom is transmitted through personal exchanges between mentor and disciple. I have called Buddhism a religion of wisdom and compassion. From the Buddhist viewpoint, what you term mercy is a manifestation in human relations of the operations of cosmological compassion. You have explained tolerance in the context of Christian virtue. I would now like to think about the Buddhist interpretation of tolerance and the relationship between Buddhist wisdom and compassion with tolerance.

Whereas Christianity is founded on belief in one omnipotent Creator God, Buddhism is based on the fundamental Law of the universe, to which Shakyamuni was enlightened. The Lotus Sutra teaches that the Law of that enlightenment transcends time, which is personified as the eternal Buddha. The Law and Buddha are one and indivisible and reveal the nature of cosmic life to be eternal, without beginning and without end. The eternal Buddha is endowed with the three kinds of 'body', or virtues: the Dharma body, the reward body and the manifested body. The Dharma body is the eternal, fundamental Law, or truth of the universe. The reward body is the cosmic wisdom inherent in and manifest by the truth. The manifested body is cosmic compassion to save all living beings. In other words, the Lotus Sutra characterizes the life force of the universe in terms of the three virtues – truth, wisdom and compassion.

Human beings manifesting these cosmic virtues in their lives are called 'envoys of the thus come one' (another name for Buddha). The following passage can be found in the Lotus Sutra:

> If one of these good men or good women in the time after I
> [Shakyamuni] have passed into extinction is able to secretly
> expound the Lotus Sutra to one person, even one phrase of it,
> then you should know that he or she is the envoy of the thus
> come one. He has been dispatched by the thus come one and
> carries out the thus come one's work.[2]

These envoys of the thus come one, sharing his eagerness for the salvation of living beings and acting on his behalf, are also called the teachers of the Law or Bodhisattvas of the Earth (or, as described in the Lotus Sutra, bodhisattvas emerging from beneath the earth).

From the Buddhist viewpoint, tolerance is a manifestation of the everlasting compassion inherent in eternal cosmic life and the eternal Buddha. In more concrete terms, it is the virtue apparent in the work of the dharma teachers of the law and Bodhisattvas of the Earth. It is what you call the virtue of coexistence. Because in Buddhism wisdom and compassion are one, naturally Buddhist tolerance is one with the wisdom inherent in the cosmic force of life.

Unger: How does Buddhism interpret forgiveness?

Ikeda: Forgiveness too is interpreted as a realization of the compassion inherent in the eternal Buddha nature. In connection with forgiveness, I am reminded of actions taken by Ceylon (now Sri Lanka) at the end of World War II. At the San Francisco Peace Conference, held in 1951, it renounced all rights to compensation from Japan.

Representatives from Ceylon expressed the opinion that, though they had the right to do so, they did not intend to demand compensation because of their belief in the words of the Buddha, whose teaching has ennobled the lives of a countless number of people in Asia, 'For hate is not conquered by hate: hate is conquered by love. This is a law eternal.'³ This passage from *The Dhammapada* (Path of Perfection) gives positive expression to forgiveness and positive tolerance founded on the cosmic virtue of great compassion.

Unger: As a Christian, I consider mutual tolerance among people of different religious backgrounds to be the working of the divine virtue of love.

Ikeda: In contrast to the good that promotes tolerance and coexistence, evil is that which causes division – a pathological condition in which people close their eyes to the things we have in common and become obsessed with differences. Going beyond the individual level, it can become the essence of group egoism taking the form of exclusivist, destructive racism and nationalism. Those who strive to overcome the lesser self, becoming enlightened to the greater, universal self by evoking the ultimate good inherent in the life of both the self and the other are called bodhisattvas in Buddhism. They put into practice positive Buddhist tolerance.

Unger: You yourself embody Bodhisattva practice.

Birth, Ageing, Illness and Death

Unger: Religion is for me the essence of humanity. How have I as a doctor in our times reached this conclusion? Are not people turning away from the great churches in growing numbers, and is this not reasonable? Certainly the answers to both questions are yes. But, as a doctor, I observe that, in difficult times, many patients cling to their religious beliefs.

Ikeda: That is an important observation. The suffering of illness makes people confront their own mortality and inspires profound re-examinations of life. Sincere concentration on life and death awakens in the mind the need for an unshakeable philosophy and religion. Nichiren teaches: 'Illness gives rise to the resolve to attain the way.'[4] He also writes: 'I should first of all learn about death, and then about other things.'[5]

The Japanese attach great importance to the appearance of the dead. Some people die with a gentle, beautiful facial expression, which is taken as evidence of their having led fulfilled and valuable lives. Others may die with any of various other expressions. How does this emphasis on final appearance strike you as a doctor?

Unger: As you say, people die with different kinds of expressions on their faces. Those whose lives have been very fulfilled and happy and those who complete their lives with confidence and mental calm can be said to die in a good state of life. And, if reincarnated, such people will be reborn in a good state. Investigations show that people whose final facial expression is distorted and frightful have met death with problems on their minds. Perhaps they lacked self-confidence and fulfilment during life.

Ikeda: Buddhism presents the solemn facts of birth, ageing, illness and death in terms of the teachings of the four sufferings. No medical advances and no amount of material wealth can solve the problems they pose. The fundamental issue of our existence is overcoming these sufferings in such a way as to maximize the radiance of life.

The story of the four encounters outside the confines of his father's palace is a metaphor for how Shakyamuni became aware of the four sufferings. It is said that on one occasion, Prince Siddhartha, the young Shakyamuni, went out from the palace by the east gate and met an old person. On another day, upon going out by the south gate, he encountered an ill person. On yet another day, when he left the palace by the west gate, he came upon a corpse. Then, on the fourth occasion, leaving by the north gate, he encountered a religious ascetic. These four encounters inspired Shakyamuni to abandon secular life for the life of religious pursuits.

Fundamentally religions exist to enable people to confront the reality of life in the form of birth, ageing, illness and death, and then to pursue the quest for true happiness. In this sense, all human beings are innately religious. As you put it, religion 'is the essence of humanity'. Turning away from the four sufferings and losing sight of the religious wisdom with which to explore the truth of life aids and abets the tendency to undervalue life.

Unger: I agree. Today issues like abortion and genetic manipulation blur definitions of the beginning and ending of human life. This in turn makes it easy for science to take advantage of life. The nature of our times makes it imperative to employ interfaith dialogues to build a culture of respect for life. On the one hand, our world is dominated by utilitarian values – that is, constant technological advance. On the other hand, we are entering an age of emphasis on life values immeasurable in utilitarian terms. Profound religious experiences are becoming increasingly important to an understanding of the meaning of life.

Ikeda: Very true. Confronting the four sufferings teaches us how to deepen our understanding of the meaning of living, and reaffirming the dignity of life makes our existence more radiant. This is the essential mission of religion. Goethe, who himself overcame serious illness, wrote in *Hermann and Dorothea*, 'The affecting image of death does not stand as a terror before the wise man, nor as the closing scene of existence before the pious man; but impels the one back into life and activity, and gives the other a comfortable prospect of escape from adversity: to both, then, death becomes *life*!'[6]

Unger: A very significant passage.

Ikeda: In a similar connection, the International Conference on Palliative Treatments, sponsored in November 2004 by the Pontifical Council for Health, Pastoral Care, and Intervention, dealt with the importance of ameliorating both actual physical suffering in patients with stubborn illnesses and the suffering caused ultimately by death itself.

A doctor of my acquaintance told me of terminally ill cancer patients who, in spite of their own suffering, worry about helping and encouraging their friends and others. As the doctor said, such people are strong in the face of illness and death. In contrast, when a crisis comes, people who have lived for nothing but their own egoistic needs prove weak. Strength is to be found in lives lived in open solidarity with others and doing everything possible for their sakes.

Compassion, as taught in Mahayana Buddhism, is the practice of eliminating the suffering of others and imparting joy. This practice results not only in the service of others but also draws forth one's own immense life force. When a person perseveres in the bodhisattva way, creative energies are summoned from within and transform the four sufferings of birth, old age, sickness and death into joy. Nichiren therefore wrote that 'we use the aspects of birth, ageing, sickness and death to adorn the towers that are our bodies'.[7] He introduced the view of life and death expounded in Mahayana Buddhism, which enables one to find joy and equanimity in death as well as in life.

Unger: Christian love is also the source of extremely positive energy – the highest virtue. Love creates perfect energy in our lives. For this reason, Jesus Christ continued to preach love.

Religion Heals the Individual and Society

Unger: To rectify the distortions arising everywhere today, we must reorganize everything to work to the advantage of humanity. When science loses sight of the fundamental need to serve those advantages, it can become domineering. The same is true of religion. When used as tools to promote the good of humanity and life, however, science and knowledge become an art.

Ikeda: That is beautifully expressed. I agree entirely.

Unger: Eugene Biser, theologian and dean of our academy, interprets both Buddhism and Christianity in a therapeutic dimension open to both individuals and humanity as a whole.

Ikeda: Shakyamuni's teaching of the four noble truths represents a therapeutic method of redressing human sufferings. They are the truth of suffering, the truth of the origin of suffering, the truth of the cessation of suffering, and the truth of the path to the cessation of suffering. Because, from the time of Shakyamuni himself, it has been entirely concerned with eliminating suffering, Buddhism can be called therapy on all levels, expanding naturally from the individual to society as a whole.

Unger: Today as society globalizes, the socially therapeutic dimension seems to be growing increasingly important. Society must deal with a large number of traditions, viewpoints and values that, within the global context, interact side by side, albeit incoherently. Compelled to cobble together their own private thought systems from what is available, people make use of all apparently suitable elements, no matter whether or not they fit together.

Ikeda: Human beings, unable to tolerate a spiritual vacuum, are constantly in search of something to believe in – a mainstay of spiritual satisfaction. What they believe in has a determining effect. That is why it is of the utmost importance to cultivate as widely as possible universally shared concepts of tolerance, compassion, love, and respect for human dignity. Losing sight of these things and lapsing into a vacuum can even threaten to lay the groundwork for new kinds of totalitarianism.

How do Europeans today relate to the extreme intolerance of totalitarian systems like fascism and Stalinism?

Unger: The extraordinary brutality and intolerance of Nazism and Stalinism and their, until then, unimaginable contempt for humanity remain a trauma for Europeans to this day. We still ask ourselves how an ideology could have swept away a highly diversified culture like ours and could have led to indiscriminate mass murder. Or, in harsher terms, what went wrong in the highly diversified European culture to provide grounds for such ideologies to operate and flourish? To be sure, dictatorial socialism was ultimately eradicated. But does this automatically mean the end of all ideologies contemptuous of humanity – the end of all hegemonic powers, which do so little for human life? Not so.

Ikeda: Totalitarian ideologies contemptuous of humanity may appear over and over in various forms. We all know that Nazism adeptly rose to power within the framework of the democratic Weimar Republic. One thing I would like to note is that behind the growth in influence of totalitarian ideologies exist many people who sympathize and empower them.

I remember that Dr Toynbee regarded fascism and communism as modern religion, which emerged to replace Christianity. But because they were in themselves too inferior for people to worship, they in fact became religions worshipping the collective power of a group.[8] I tend to believe this observation is valid.

Unger: Japan, too, has experienced similar traumas. How do Japanese artists and intellectuals discuss the past, and what conclusions do they reach?

Ikeda: Very few artists, intellectuals and religious leaders put up any resistance to rampant, contemptuous totalitarianism. Josei Toda resisted totalitarianism based on State Shinto as its spiritual support so intensely that he was imprisoned. This is one of the reasons why I put absolute trust in him. It is true that in postwar Japan, when professions of pacifism and antiwar sentiments ran high, some artists and intellectuals sincerely and earnestly opposed totalitarianism. At the same time, however, many of them avoided directly coming to terms with the sheer misery Japan caused other Asian nations. This tendency continues to this day.

The political philosopher Hannah Arendt (1906–75) has used the term 'hole of oblivion' to express the way that totalitarianism erases popular memory. Many Japanese have consigned the misery Japan inflicted on others to such a 'hole'. The resulting failure to overcome the totalitarian mindset fully is one of the reasons why the citizens of the Asian countries Japan invaded still mistrust us.

Unger: The reality the media reports to us daily from the whole world terrifies me. It is really ghastly. Intolerance is the bad in us human beings. But I say again that tolerance is in our genes. In spite of the horrors of present global events, I view possible future developments optimistically. Two factors make me think this way. One is to be found in the tradition of Joachim de Fiore (1130/35–1201/2),

Nicolaus Cusanus (1401–64), and Pierre Teilhard de Chardin (1881–1955). All three of these thinkers were convinced that the human spirit develops and that we are capable of spiritual maturity. Teilhard offered perspectives on genetic engineering in this connection. Genetic technology shows us that we are all the result of variation, thus the result of mutations. That is why I entertain the justifiable hope that humanity can develop in ways that will ultimately make peaceful coexistence possible.

Ikeda: All three of the philosophers you mention believed that the human spirit can grow infinitely. This belief gave them hope for the future. Nichiren, too, had radiant hope for peaceful symbiosis among members of the human race. He lived in thirteenth-century Japan at a time of widespread pessimism and despair. In Buddhist history, it was what is called the Latter Day of the Law, a time 2,000 years after the death of Shakyamuni when Buddhism falls into decline, human life becomes increasingly defiled, and the society and age are confused and corrupt. In fact, Japan as a whole experienced a series of fires, floods and earthquakes as well as epidemics and famine. But Nichiren declared 'When great evil occurs, great good follows'.[9] He believed that precisely because of these mounting crises, it meant the time had come to create from confusion a new spiritual flourishing and to realize a world of lasting peace. He saw himself as the active leader of this historic reformation.

Unger: Fascinating. The second factor sustaining my optimism is the evolution of Europe from fractured bellicosity to an increasingly apparent unity. In this process, we always confront the problem of how to overcome old nationalism and forge European unity while at the same time preserving vital regional structures within the total framework. Europe has achieved two striking, encouraging things: the Schengen Agreement and the euro as the uniform currency for eight countries. The Schengen Agreement has done away with border controls in many – though not yet all – European countries.

Ikeda: By preserving local regionality, Europe voluntarily moves towards unity in a peaceful way. Its pluralistic, nonmilitary order is expected to serve as a model for a new regional and global order. I am reminded of the Austrian Count Richard Coudenhove-Kalergi (1894–1972), who advocated the Pan-European concept. He and I began to engage in dialogues from 1967. At the time, I was young enough to have been his son. His earnest way of speaking made me feel as if he were bequeathing a mission to me. Our dialogues were published as the book *Bunmei nishi to higashi* [tentatively translated as Civilization, East and West]. In it, Count Coudenhove-Kalergi said that averting a third world war will be possible only if we consistently stress the importance of human symbiosis and mutual trust on the basis of a spiritual movement transcending conflicts arising from racial, religious, ideological and national differences.[10] This statement is more important now than it was when originally made nearly four decades ago. A federation of compassion and tolerance which creates bonds of mutual trust is essential if we are to halt the steadily increasing violence in the world.

Unger: I became personally acquainted with Count Richard Coudenhove-Kalergi at a Pan-Europe congress in Vienna in 1964. He was descended from the Habsburgs, who, guided by the principle of live and let live, ruled Central Europe in an exemplary fashion. Many languages were brought together under one crown; that is, one idea. The tradition, which goes back to Charlemagne and the ruling ideas of the Holy Roman Empire, provided a bond assuring people that they could live well. Today we need a similar European idea operative in more than the fields of bureaucracy and economy. In 1923, Count Coudenhove-Kalergi took up this idea and developed a brilliant vision from it.

Ikeda: Yes, *Pan-Europe*, which Count Coudenhove-Kalergi published and dedicated to the youth of Europe, reads in part:

> Whether an idea remains a utopia or becomes a reality usually depends upon the number and the energy of its supporters. While thousands believe in Pan-Europe, it is a utopia; so long as millions believe in it, it is a program; but at once a hundred million believe in it, it becomes a fact.[11]

How delighted he would be if he could see that hundreds of millions today believe in his vision for a Pan-Europe.

Freedom Central to the European Idea

Ikeda: The Danube Union, a source of Count Coudenhove-Kalergi's Pan-Europe concept, was Austrian in origin. During its long history, Austria has evolved profound wisdom in connection with tolerance. The Habsburgs tried to incorporate the spirit of tolerance in governance. Historically their empire uniquely served the purposes of a pan-European communality and in various ways strove to have all the ethnic groups and languages under its rule recognized as equals. For example, Article 19 of the Austrian Constitution of 1867 guaranteed equality of all nationalities under the monarchy and declared that all peoples had the inviolable right to preserve their nationality and languages. It can thus be said that Pan-Europe is irrevocably tied to the Habsburgs and Christianity.

Unger: The imperial tradition of Charlemagne, which lasted for a thousand years, until the time of Charles I (1887–1922), would have been impossible without a humane concept of empire. Even today there exists an idea of Europe under which everybody can live in freedom. One of the ingredients of Christianity, freedom is based on and expressed in the Gospels, especially the Epistles of Paul, where it is said that human beings can develop only in conditions of spiritual freedom. But, in this context, when put into action, freedom is to be understood as paired with the individual conscience.

Ikeda: Exactly. Otherwise, freedom can lapse into self-indulgence. That is why religious cultivation of the conscience is essential in a free society.

Fall of the Wall and European Revival

Ikeda: During the Cold War, a split between the East and the West ruptured the spirit of symbiosis cultivated over long years. But the European world started reviving when the Cold War ended.

What were your reactions to the fall of the Berlin Wall some fifteen years ago?

Unger: 9 November 1989, the day the wall came down, had tremendous significance for Europe. It is called *Wende* (the turnaround) because it totally changed the lives of Europeans. World War II ended then, in the truest sense of the word. It was a very moving experience and meant that relations among Central European nations were being normalized. My wife Monica and I went straight to Berlin, where we hammered off a bit of the wall, which we still have at home. No one had imagined that communism would collapse all at once without a single gunshot. I felt certain that the destruction of the wall opened new, hopeful perspectives for Europe.

Not long ago, I travelled to Riga, the capital of the Republic of Latvia, where I clearly stated that, in the past, there had never been problems between Russia and Europe. The trouble started when communist ideology split Europe apart. That is all over now. From the European standpoint, discourse with Moscow or Saint Petersburg is nothing special. Long ago Russia, Germany and Austria talked things over in a natural set of relations that was, unfortunately, discontinued for about a century.

Ikeda: I visited Berlin in October 1961, two months after the Berlin Wall went up. To some friends who were with me I said, 'I am sure that in thirty years, this Berlin Wall will no longer be standing'. I was expressing not a wish or a prediction, but a deep belief that the freedom- and peace-loving human spirit would triumph. At the same time, I was expressing determination to devote myself entirely to making my mentor Josei Toda's dream of global citizenship a reality.

Unger: I believe that, as Josei Toda said, we must all arrive at an awareness of being citizens of the Earth, in spite of speaking different languages, living in different cultures, and entertaining different religious views. This idea ought to unite us instead of driving us apart.

Ikeda: We must cultivate among the peoples of the world a spiritual solidarity based on awareness of our shared global citizenship and in this way provide power for overcoming all kinds of conflicts, including those among nations.

Limiting Freedom of Expression

Unger: I would like to follow our general discussion of tolerance with related issues on the social plane. For example, in our current information culture, violence portrayed in the mass media seems to encourage intolerance. Do you agree?

Ikeda: Freedom of speech and expression is absolutely indispensable to the building of a culture of peace and tolerance. Many people, however, question the purpose of such freedom when its name is used by the mass media to threaten human rights through licence and irresponsible commercialism. The tendency to complain about this seems to justify authorities' attempts to restrict freedom. In other words, the media themselves endanger the foundations of liberty.

Journalism can fan the flames of racial hatred and xenophobia, thus directly threatening the emergence of a tolerant culture. We must remember the Nazi dictum that people will believe any lie as long as you repeat it to them often enough. History offers many instances in which lies have fostered increasing hatred. Lies open the way to violence. When they trample on human rights, lies themselves are violence. This is why repudiating lies and safeguarding the truth should be the ethos of a tolerant culture.

The famous American economist John Kenneth Galbraith (1908–2006) shared with me his belief that journalists have responsibilities that they are obliged to uphold as human beings, the main one being truth. And while he felt the need for some laws to restrain those reporters who do not resort to the truth, he insisted that such sanctions should only be used after careful scrutiny.[12]

Unger: Determining what kinds of laws are required to protect tolerance is a major issue. Without social limitations, people who stir up racial hatred will continue sowing their seeds until tolerance is obliterated. On the other hand, political correction by thought police creates only superficial restrictions on expression. Underlying hatreds build up and are certain to explode sooner or later.

Artists should be permitted forms of expression that provoke shock. But at what stage should an artist be informed that society can tolerate no further provocation? Balance must be maintained, but it always entails risk. Too much lenience can destroy the social fabric. In the wonderful parable play *Biedermann und die Brandstifter* (The Firebugs) by the Swiss writer Max Frisch (1911–91), the character Biedermann is so lenient toward arsonists that he even provides them with matches. I believe we can maintain the needed balance as long as we take pride in our own value.

Ikeda: Freedom of expression is the basis of liberty and democracy. Though the extent to which nations and communities allow for freedom of expression varies with cultural conditions, restrictions on it should be minimized. Maintaining a balance between freedom of expression and limiting expressions of violence, hatred and discrimination requires a holistic and positive approach, including both the legal system, self-regulation and education.

Education is fundamental because it elevates the standards of both those who transmit and those who receive media information. In more concrete terms, media literacy – the ability to discriminate, evaluate, and apply media information – must be thoroughly improved. Education that achieves these ends in the home, the school and the community endows the general public with the autonomy to use and criticize the media independently. This is the best way to improve the media. Education should encourage people to regard the media in the spirit of critical, independent dialogue, thus preparing the ground for a culture of tolerance and peace.

Research into the effects of the media apparently recognizes a correlation between violent expression and actual social violence without going so far as to establish clear causal relations. Of course, excessively violent and provocative media content cannot be conducive to wholesome psychological development in children.

Unger: I cannot understand why experts go on insisting that there is no connection between media violence and aggression among young people. Ultimately television is largely financed by advertising. The advertiser wants his products and services to sell. Advertising on television patently has an effect; violence is presumed to have none. This incredible hypocrisy in our society must be addressed in the name of tolerance.

In April 2002, in the small town of Erfurt in Germany occurred a crime of the kind that, for a long time, we were familiar with only from the United States. A nineteen-year-old expelled student went heavily armed to his former school and shot fourteen teachers, two students and another adult. He stopped only when a teacher who met the assailant in the hall ordered him to remove the mask he was wearing and look him in the eyes.

In this instance, I see two things as clearly as if they were under a magnifying glass. On the one hand is the heartless bloody work of a young weapons maniac who spent his time watching violent videos and playing violent computer games. On the other is a shining humaneness capable of influencing a soul as brutalized as that of the young killer.

Ikeda: We must always believe that the spirit of Buddhist compassion and that of Christian love can change anyone. The conviction that this is true supports my abiding insistence on dialogue before armament. I am resolved to join you, a chevalier of love and tolerance, in the continuing struggle for peace instead of war, civilization instead of barbarism, and tolerance instead of intolerance.

CHAPTER THREE

Creating a Culture of Peace

————————

Ikeda: To reverse trends that made the twentieth century a century of war and violence and make the twenty-first century one of peace and non-violence we must effect a paradigm change. We must transform the culture of war into a culture of peace, thus altering the grounds from which conflict and violence emerge worldwide. How are we to do this and what conditions are required? To answer these questions, taking into consideration and consolidating what we have already said about tolerance and compassion, I would like to examine three major topics: human rights as the indispensable element in a culture of peace, the role of women as peacemakers, and humanistic education as the key to peace.

Unger: I agree on the urgent need to create a culture of peace. But I am very disturbed by the dangerous and disgusting way our world is developing. Rapid globalization is crushing ethnic and regional cultures. Economic crises brought on by the discrepancy between the prosperous industrialized nations and struggling developing nations spark sudden outbreaks of ethnic war and slaughter. We have only to read the newspapers to realize at once that war is at our doors. Moreover, we now confront terrorism, the roots of which are traceable to unresolved ethnic conflicts. We witness how a mixture of religious and ethnic fundamentalisms leads to violent terrorist acts.

Ikeda: After the attacks on the United States on 11 September 2001, fear of terrorism spread as tension rose throughout the world. Political, cultural and economic globalization transcending national and ethnic boundaries aims for unification. Reaction against it, however, leads back to nationalism and fundamentalism, which is the source of terror-related extremism. At the same time, owing to the rapid expansion of global transportation and communication networks and the consolidating effect of multinational corporations, we can come into contact with the traditions, philosophies and value systems of diverse historical and cultural backgrounds from all over the world. Whereas this reception of diverse values can enrich our thought, it also destroys cultures and upsets value systems. In this sense, our age of global integration should not become an age of confusion.

Today we are in a state of chaotic conflict between centripetal standardization and centrifugal disintegration. While continuing to seek universal values, we must respect and accurately appreciate originality and individuality. Here again, balance is essential. Obsession with particularities and isolation from prevailing contexts, traditions and value systems breeds self-righteousness, prejudice and discrimination. Single-minded pursuit of universality, on the one hand, runs the danger of global standardization; exclusive emphasis on individuality, on the other, threatens global break-up.

Unger: You pinpoint many of the things that worry me greatly. I observe with concern the way fascist nationalist movements stimulate unacceptable things that contribute nothing to global understanding. Too many people remain indifferent to the future of humanity and the world. Such indifference leads to the abuse of human rights and undermining of freedom.

Ikeda: Economic disparity, oppression and discrimination create hot-beds of terrorism. Indifference to them amounts to the inability to imagine others' sufferings. This reflects in the modern malady of failing to acknowledge the existence of other people. The tendency to be uninterested in society and others and to lapse into self-absorption is especially apparent in young people, who bear the responsibility for the future.

Unger: It seems to me that, in addition to indifference, a sense of isolation and aggression spreading among the younger generations causes lapses into individual alienation and extreme individualism. On the basis of all kinds of information made available by globalization, the young become attached to a materialistic lifestyle. Often they value money above human life. Such ultra-individualism is a threat to our world.

Ikeda: Together with materialism, this trend toward ultra-individualism or selfishness is another modern malady. The selfish are enclosed in their own small worlds and have no interest in the pain, suffering, and sorrow of others. Trifles arouse their strong emotions and launch them on violent acts.

Exclusive selfish concern with satisfying one's own desires serves as a breeding ground for conflict by facilitating inequality, prejudice and discrimination.

A Parable

Ikeda: My own image of a culture of peace is one in which all living beings can manifest their distinct characteristics while existing in a state of equality. How can we transform the aggressive and destructive culture of war into a culture of peace through dialogue and cooperation?

I would like to mention an image described in the parable of the three kinds of medicinal herbs and the two kinds of trees in 'The

Parable of the Medicinal Herbs' chapter of the Lotus Sutra, which portrays an ideal image of peace in Buddhism. The text describes how all kinds of trees and herbs with their different names and characteristics grow together. 'The rain falling from one blanket of cloud accords with each particular species and nature, causing it to sprout and mature, to blossom and bear fruit.'[1]

The plants symbolize human culture. The image is one in which the various herbs and trees, though diverse in nature, grow from an impartial, enriching rain and earth – a symbol of a benevolent universe preserving their individual traits as they thrive and flourish in a culture of peace.

Unger: I see how the image of coexistence, diversity and equality in the parable you refer to indicates a way to establish a culture of peace. In the present millennium, all humanity must address the problem of collaborative development. In doing so, we must realize that the lives of people of all ethnic and religious backgrounds are equally important.

Ikeda: Respect for diversity, dialogue, and insight into universality are requisite conditions for the creation of a culture of peace. Respect for diversity means that we recognize and honour people different from ourselves, learn from them, and assimilate what we learn. With this kind of open-minded attitude, individuals, groups, and nations can constantly develop and improve. The closed mind that excludes others, on the other hand, consequently diminishes itself.

Unger: I agree. For a culture of peace, we must respect the diversity of life and nature and create universal human values. This requires tolerance and morality. Tolerance is the indispensable bridge to a new world where people mutually recognize the dignity of life in all their encounters.

Ikeda: Only a culture of peace can direct us toward a global civilization of abundance in which diverse cultures stimulate each other in significant ways. As we have said, the spirit of active tolerance is indispensable.

Unger: Indeed. The peoples of the world – Buddhist, Christian, Muslim, atheist – must realize the important value of active tolerance.

Ikeda: Throughout history, civilizational, cultural and ethnic discrimination has bred innumerable tragedies that the spirit of active tolerance might have averted. We need to respect other cultural views of the world and of ethics and humbly learn from them.

Unger: That is exactly what I think. In 2004, the European Academy of Sciences and Arts established in Salzburg the Cardinal König Institute as a forum for religious believers and non-believers alike. (The cardinal was honorary president, until his death not long after the institute was founded.) Its goals are to conduct inter-religious dialogues, promote exchanges of opinion with atheists, and convert intolerance into tolerance. In your words, it is a place to respect other cultural views of the world and of ethics and to learn from them.

Ikeda: I had great respect for the perceptiveness and achievements of Cardinal König. Dialogue, my second condition for a culture of peace, is a practical way to cultivate the spirit of tolerance. The culture of war gives precedence to perceptions of one's own cultural supremacy and is closed to all others; the culture of peace, on the other hand, is open to other cultures.

Achieving openness is not always an easy process. As Arnold Toynbee wrote in *The World and the West*, 'The reception of a foreign culture is a painful as well as a hazardous undertaking'.[2] While it would be naive to expect them always to proceed smoothly, such encounters are not necessarily fated to end in distorting, destructive clashes.

Count Richard Coudenhove-Kalergi once told me that most great cultures have been built on things absorbed from other cultures. Dialogue is the practical way to be open to other cultures.

Unger: Abrahamic religions have stressed the importance of dialogue, as is especially apparent in the Christian New Testament. In a speech I delivered at Soka University in July 1997, I told the students that our ability to engage in dialogue will determine the further fate of this Earth.

Dialogue generates the virtue of tolerance. Learning from and acting in accordance with that virtue lay the foundation for peace. Interfaith dialogues bridge philosophical gaps and create the basis for global peace.

In a wonderful way, through dialogue, you not only set forth your own beliefs, but also attentively heed the voices of people from other religious and cultural backgrounds. All of their ideas and pronouncements contribute to the realization of world peace. Moreover, I am profoundly moved by the way you encourage your many friends to keep up the struggle to live creative and valuable lives.

Ikeda: Thank you for your kind words. I am convinced that dialogue has the power to transcend differences and unite our world. On the strength of this conviction, I have engaged in intercultural and interfaith dialogues with intellectuals and leaders from all over the globe.

Unger: Yes, you have taken action. To make contacts, instead of waiting for others to arrive, it is necessary to advance and meet them in a positive manner while, at the same time, having firm opinions of one's own.

Ikeda: I agree. Engaging in dialogue is not the same as simply listening to whatever others have to say. Dialogue builds mutual understanding and trust. Without firm faith and philosophy, it is impossible to understand other people or to engage in real dialogue with them. Without dialogue, human beings wander in the darkness of self-righteousness. Dialogue shines light on that cold darkness, showing us the path we ought to follow.

Insight into the Universal – the Eternal Value of Life

Ikeda: Ultimately eternal values – insight into which is my third condition for a culture of peace – emerge on the expansive plain illuminated by dialogue. Extensive exchanges or even cohabitation with other cultures provides deeper insights into one's own culture. Delving into the inner nature of a culture or tradition constitutes a return to the sublime spiritual sources from which they spring and makes possible the discovery of universal values among particularities. Universal values rooted in the eternal underlie the spiritual energy that formulates all great cultures and civilizations.

Ethics based on such universal values should be formed from interaction with modern civilization and put into practical application. Accomplishing this is the task of people of religion. A religion that views the individual as an entirety and seeks harmony with society and the natural environment plays the important role of pursuing universal values while permitting the manifestation of individual originality.

Unger: Religion has the great task of guarding values. When it is founded on the eternal value of life, it also serves as the nucleus of culture formation. By accepting its mission in connection with eternal values, it serves as a basis for a humanity-wide culture and contributes to the building of cultural tolerance. Human beings can spontaneously contribute to culture only when they revere life and respect all of God's creation. I deeply respect the way in which, on the basis of the value of life, the expansive network of SGI engages in mighty, moral activity for the building of a culture of peace. Anyone who strives for peace must agree with the operations of your organization.

Ikeda: Religions in the twenty-first century require the magnanimity to transcend sectarianism and cooperate in contributing to the happiness of humanity. For the sake of peace, they must plumb the depths of their own universal values of life and conduct repeated dialogues in the spirit of global citizenship. This I believe is the role of people with religious faith today.

The concrete task is to cultivate the spirit of positive tolerance in the mind of each individual. Since the inception of SGI, its members have combined their own individual human revolutions with effort for the betterment of society. We are resolved to persevere in this course of action steadily and surely in the years to come.

Unger: When we strive to live together in harmony and spontaneously and freely adopt tolerance as a central tenet we are able to see all humanity with an undeviating eye. Beginning with the individual, like spreading fire, this will transmit itself to the family, the community, the nation, and across continents. Taking the first step toward peace from one's own home initiates the process of creating conditions for expanding tolerance and virtue. That is why I expect great things from the further development of SGI.

Desire for Peace Expressed in the Universal Declaration of Human Rights

Ikeda: The strict observance of human rights is indispensable to establishing a durable culture of peace. Nonetheless, all over the world today discrimination and other violations of human rights are widespread, as is especially apparent in the complex of issues including terrorism, internecine war, poverty, hunger, and HIV-AIDS plaguing Africa.

Unger: The environmental destruction, poverty and war that are taking place in Africa are one of humanity's greatest tragedies. Much of the responsibility for it falls on the great Western powers and their colonizing policies. As a physician dedicated to protecting life, I am immensely pained by the misery of the current African situation.

Ikeda: Aid to Africa was one of the major topics of the G-8 summit meeting held in Scotland in 2005. Twelve million African children are reported to have lost one or both parents to HIV-AIDS. Out on the streets, these so-called AIDS orphans live in worsening poverty. They are deprived of food and clothing, sold into slavery or military service, and forced into prostitution or beggary. Eliminating the poverty that causes this misery is essential to ensuring the observation of human rights in the twenty-first century.

Unger: Anyone lacking the passion to protect life also lacks passion for human rights. Human rights are one of the most meaningful norms won at a heavy price. The history of human rights arises from European secularization. It must be remembered that at one time the Church totally dominated daily life. Secularization originated with the French Revolution, and the concept of human rights is its grandchild. The notion of human rights came into being in the European context and it greatly facilitated European democratization in the twentieth century.

Do you think human rights are universally applicable and part of a non-European global tradition or do they reflect largely the European tradition, especially the tradition that started with the French Revolution?

Ikeda: Undeniably Europe has played a major role in establishing human rights. For instance, the eighteenth-century European Enlightenment was a mainstay in the struggle for human rights proclaimed in the American and French revolutions. In the twentieth century, in Europe and many other parts of the world, people strove to create a firm foundation for the universalization of respect for human rights. The culmination of these efforts was the Universal Declaration of Human Rights adopted by the third meeting of the United Nations General Assembly in 1948. In the declaration we can perceive a firm resolution that the atrocious violation of human rights committed during World War II should not be repeated.

Among the people who made major contributions to the compilation of the Universal Declaration are Austregésilo de Athayde (1898–1993), former president of the Brazilian Academy of Letters – with whom I engaged in a dialogue on the topic of human rights in the twenty-first century; Dr René Cassin (1887–1976), former president of the European Court of Human Rights at Strasbourg and laureate of the Nobel Peace Prize; and Eleanor Roosevelt (1884–1962).

Mr Athayde gave precedence to human rights over political systems and insisted that, eternal and universal, they must be free of national or historical restrictions.[3] In the search for a correct substantiation for their foundations, he and his co-drafters of the Declaration traced the history of concern with human rights all the way back to the Code of Hammurabi from the eighteenth century BCE.[4]

The Sacred within Humanity

Ikeda: In our discussion of the universal basis for human rights, Athayde mentioned the Mosaic Decalogue. In response, I mentioned the five precepts and the ten good precepts, both of which are basic ethics to be observed by Buddhists. They both begin with the prohibition against destroying life. Put differently, they call for non-violence and respect for life. Buddhism stresses the dignity of life because each individual life is endowed with the Buddha nature and with the possibility of manifesting it.

As Athayde said emphatically, 'I sympathise with this Buddhist view because I am convinced that appreciation for the dignity of humanity cannot gain wide acceptance unless we become aware of the sacred element within ourselves'.[5] And he stated that Buddhism has its own version of the evolution of human rights.[6]

Nichiren clearly proclaimed humankind's innate freedom. Living in feudal Japan, he suffered oppression at the hands of rabid state authorities. But he declared, 'Even if it seems that, because I was born in the ruler's domain, I follow him in my actions, I will never follow him in my heart'.[7] These words represent a crystallization of the philosophy extolling the freedom of the human spirit. They are included in *Birthright of Man*, compiled by UNESCO to commemorate the twentieth anniversary of the Universal Declaration of Human Rights.

Honouring human rights expands the world of tolerance, whereas despising them allows the evil of intolerance to run rife. This is why the twenty-first century must witness redoubled efforts among world religions, including Christianity and Buddhism, to bring forth from their traditions of thought and spirituality a philosophical basis for human rights.

Unger: During the twentieth century the world was subjected twice to global conflict. To prevent the recurrence of this tragic history, we must passionately declare the value of human life and guard the dignity of God's wonderful creation, humankind. In the past century, the concept of human rights was used by the West to restrain the nations of the Eastern bloc. But an honest examination of their own records shows that Western nations too often infringe human rights. We must break from this political framework and enrich and expand the idea of human rights itself. In this age of globalization, we must adopt a new viewpoint incorporating tolerance and respect for the dignity of life. What are your views on future developments in human rights?

Ikeda: The interpretation of human rights is changing greatly because global environmental problems and the damage of war generate issues not covered by the concept as understood in the past. This results in what is called the third generation of human rights extending to rights related to the environment, peace, growth and development. The expansion is intended to embrace the rights of people in the future and of plants and non-human animals as well.

The enrichment and expansion of the concept of human rights requires the spirit of tolerance and respect for the dignity of life. We also need a firm philosophy further clarifying the dignity of life. We must replace emphasis on corporate and national interests with emphasis on the interests of humanity. Solving human rights related problems now demands global popular solidarity. We must all join together to overcome national, ethnic and corporate egoism.

Unger: I agree. The ultimate goal of humanity is world peace and improvement of the global environment. The issue of human rights is a bridge toward intercivilizational dialogue on peaceful coexistence for all humanity. We must pay close attention to the causes of division among people, economic discrimination, and exploitation. Exerting one's own rights without taking the rights of others into consideration is egotistical, inviting environmental destruction and unhappiness. It is important to avoid such actions and to understand that the existence of others enriches the self.

Ikeda: Very true. Protecting one's own rights while those of others are being abused is unacceptable. *We*, not *I*, should be the word of the century. We must make living together, supporting each other, and mutually prospering the spirit of the age.

Unger: In that sense, we need, as it were, to shift gears in connection with human rights. Abandoning the egoism that permits exploitation, we need to cultivate and educate ourselves as human beings and learn to esteem the rights of others. Exclusive concern with ensuring one's own rights reflects a petty state of mind. Our true goal should be to develop, enlighten and vitalize ourselves. In this way, without resorting to exploitation, we can live so as to vitalize other people and all creation. Ultimately this relates to the building of a peaceful society.

Ikeda: The internal development and flowering of human nature is the goal of human rights and a central source of happiness. Harmonious coexistence is impossible as long as everyone insists solely on their own rights while ignoring those of others. Under such conditions, human rights cannot be safeguarded.

Unger: Human rights in the new millennium are still insufficiently honoured. Throughout the twentieth century, individual freedom and other rights were extremely important in that they were means to rationalize and gratify materialistic human desires. With advancing globalization, however, these rights have merely become means for the smooth conduct of our daily lives. In our age, it is enough for each individual to insist on them. At the risk of misunderstanding, I might say that human rights are only a partial element in cultivating and bringing to full flower the all-round human being.

Ikeda: The well-rounded human being has a caring eye for the conditions of other people and is capable of self-regulation. Such a person never tries to build their own happiness on others' unhappiness.

The Austrian poet Hugo von Hofmannsthal (1874–1929) once wrote that a person who truly 'lives' life identifies with others and shares their sufferings and joys.[8] He added that such a person does not remain aloof and clinical in his or her relationship with those around them.[9] The spirit of self-regulation begins to operate when we realize that we exist because others do and feel concern for our surroundings, including other people. Neglecting self-regulation and insisting on one's own rights alone destroys the environment on which we depend, thus endangering our own continued existence. This is why cultivation of a total human being capable of self-regulation and caring for others is indispensable to creating a century of human rights.

As I mentioned earlier, Nichiren wrote that 'the varied sufferings that all living beings undergo – all these are Nichiren's own sufferings'.[10] This spirit of compassion and sympathy is the wellspring of the SGI movement and should form the foundation of the human rights philosophy of the twenty-first century.

Unger: Essentially all the great religious founders – Abraham, Christ, Shakyamuni, and so on – had peaceful coexistence and the prosperity of the globe and of humanity as their goals. Their teachings are intended to promote the attainment of these aims. Unfortunately, however, religious leaders today cannot always be said to pursue them. To do so, people of religion must address major issues like peace and the global environment.

Ikeda: It is time to return to the starting point of all the great founders who, wanting to liberate humanity, sought human happiness. More than a century ago, Makiguchi insisted that humanity should go beyond military, political and economic competition and engage in humanitarian competition. People themselves must play the lead part in bringing about this change.

Dr Toynbee said that history is driven by deep, barely perceptible undercurrents. The activities of each person and their collective power create the undercurrent of an age. People aware of their own dignity and the dignity of others create change from where they are.

Revolutionizing each individual and developing whole populations of individuals capable of such initiative is the key to altering the destiny of humanity. This is the fundamental ideal of SGI.

We must wrathfully and relentlessly combat anything that threatens or tramples on the dignity of life. Anger directed against evil is good and value-productive. Nichiren wrote, 'You should understand that anger can be either a good or a bad thing'.[11]

Struggling against evil develops good in the self and the other. Respect for human rights is no abstract concept. It cannot be achieved through legislation or upheld by any system. It must emerge from the ceaseless struggle against the iniquities that trample on the dignity of life. I believe that the dawn of a century of human rights will begin when the ordinary people proudly awaken to their universal dignity, tackle actual issues and initiate humanitarian competition.

Peace Begins in the Home

Ikeda: I believe a century of women is synonymous with a century of peace. This is because women should and must play a major part in creating a culture of peace. Of course, I am opposed to strict role delineations on the basis of gender. The important thing is for both men and women to be equally happy. Assigning some roles exclusively to men and others exclusively to women brings unhappiness and represents mistaken priorities.

Unger: I expect great things from women in peace-related work. And I agree with you that assigning roles purely on a gender basis is a mistake. Essentially man and woman constitute humanity as a whole and form the foundation for the future. Holy Scripture does not adequately set forth the role of woman. Nonetheless, Christian teachings hold that God created humanity male and female and entrusted to them both self-instruction and the instruction of

others. Both sexes possess incomparable dignity and must contribute
to the value of the dignity of all life.

Ikeda: I agree. Though he lived in a time when women were largely
despised, Shakyamuni never discriminated against them. He admit-
ted them into the Order, where they continued to play an active
part even after his death. More than seven centuries ago, Nichiren
strictly warned against gender inequity, believing that both men
and women have noble missions: 'There should be no discrimination
among those who propagate the five characters of Myoho-renge-kyo
in the Latter Day of the Law, be they men or women.'[12] Surely the
most valuable system is for men and women in the home, at the
workplace and in the community to make full use of their individual
traits through mutual respect and equality.

The preamble to the Tolerance Charter of your academy laments
the current situation in which 'the family becomes endangered,
too, and is less and less able to perform its main task of being a
stable core of human communities'.[13] Personal experience has taught
me that family and friends provide the motivating power for vig-
orous social activity on the part of the individual. The home is the
basis of human relations, and women play a major part there.

Unger: Vibrant social solidarity is unachievable without the home. But
today, the family and the home are losing their power to stabilize
society and create solidarity.

Ikeda: The home is the smallest, most important social unit. Stability
in society is impossible without stability in the home. The psychologist Erich Neumann (1905–60) has written that men and women
are bound by confrontational elements: day and night, patriarchal
consciousness and matriarchal consciousness, and so on. While manifesting their distinctive productivities, these elements mutually

complement each other and help bring each other to fruition and union.[14] Through this process, it is possible to build a stable home. Children come to regard the home as a safe harbour when parents try to see things from their viewpoint and share both their troubles and joys. The optimistic, open-minded attitudes of parents toward the community and society are instilled into the minds of children and become a driving force for social cohesion.

Eleanor Roosevelt said that universal human rights begin in small places close to home.[15] Indeed, the home is where the spirit of tolerance and awareness of human rights form and flourish.

Unger: On the strength of their unique characteristics, women play the essential role of stabilizing the home. Although in recent years men have started helping, child-rearing has always been a field in which women have made the most outstanding social contribution. I often wonder why woman's wonderful work in bringing up children and stabilizing the home fails to get the recognition it deserves.

Ikeda: When I first met your family in 1997, your wife Monica gave me a glimpse of the importance of the partner's role when she said that she had been a mother for more than twenty years without ever feeling victimized. It may be that the children of mothers who feel motherhood victimizes them consider themselves victims as well.

The American futurologist Hazel Henderson, with whom I published a dialogue entitled *Planetary Citizenship*, is well known for her concept of a love economy. Women support the corporate economy by keeping house, rearing children, caring for the ill, and performing community services. Though their real productivity is half the total, their efforts are not calculated in a Gross National Product (GNP) and go without remuneration. Dr Henderson insists that this love economy ought to be adopted as an index replacing the GNP. Her theory sheds light on the social contributions of women, heretofore ignored by economics, and on the importance of caring, sharing, and prizing life and nature.

Unger: Many women today must harmonize their maternal duties with outside work. Aside from the biological role of rearing life, however, their social roles remain underdeveloped.

Ikeda: We must justly evaluate women's power and prize their opinions and roles. The Indian poet Rabindranath Tagore (1861–1941) said that women have a stronger inner vitality than do men[16] and that their strength is indispensable to the forging of a spiritual civilization.[17] In view of the contemporary male-dominated civilization based on strength, he hoped that woman-power would cultivate a civilization of the soul founded on compassion. He wrote that 'the next civilization, it is hoped, will be based not merely upon economical and political competition and exploitation but upon worldwide social co-operation; upon spiritual ideals of reciprocity, and not upon economic ideals of efficiency. And then women will have their true place.'[18]

Gandhi too clearly and directly expressed his conviction that women hold the key to the creation of a non-violent world. As he stated, 'If by strength is meant moral power, then woman is immeasurably man's superior. . . . If nonviolence is the law of our being, the future is with woman'.[19]

Women are best equipped to understand and relieve suffering because throughout history, in times of social unrest, war, violence, suppressed human rights, famine and plague, they have suffered most. A sense of responsibility for guarding the future of children lends great strength to their voices. I believe that in the twenty-first century women's activities and contributions in many fields will reform first society then the very fabric of civilization. Essentially women are pacifists emotionally endowed with the ability to protect and have compassion for life.

The Austrian writer and peace activist Bertha von Suttner (1843–1914) shows what great history-changing achievements are possible for one woman sincerely devoted to peace. Her anti-war novel *Die Waffen nieder* (Lay Down Your Arms, 1889) had a great impact on the reading public. In spite of misunderstanding and slander, she resolutely went on writing and lecturing, fostering unity for the cause of peace. Her influence on Alfred Nobel was instrumental in his establishing the Nobel Peace Prizes. When a debater taking part in a peace conference at the University of Vienna – your alma mater – argued that individuals are incapable of altering history, she energetically claimed the opposite. She was a living testament to this claim.

Of course any adult, male or female, who has a sense of responsibility for the future and wants to do something about it can become a driving force in halting the culture of war and cultivating the culture of peace.

Unger: Very true. For the sake of the general peace and prosperity and in considering the roles of men and women, we must think in terms of the whole world. Our existence is the outcome of the cooperative endeavours of both sexes. Men and women all have roles to perform. When each person plays his or her fate-allotted part they can experience personal development and triumph.

Humanistic Education

Ikeda: Humanistic education is the foundation for a culture of peace. A cause of great concern in Japan during these past years has been rising numbers of atrocious crimes perpetrated by younger and younger individuals. A tendency to undervalue life in a society concerned mainly with efficiency and materialism contributes greatly to this trend. The power of education is of great importance to achieving the imperative goal of halting the worrying trends of heinous crime and prevalent violence.

Unger: The situation in Europe is similar. As I have already said, we are experiencing powerful secularization, which marginalizes and muffles the church. People have forgotten 'Thou shalt not kill'. The value criteria shared by all religions are being lost and replaced by prevailing materialistic values. Globalization of the mass media and homogenization of the world add impetus to the trend. The outcome is disregard for the value of life and frequent killings. All of this relates to education.

Ikeda: I agree. Education enables young people, the community, and whole nations to create the future. But Japanese education today faces mountainous issues like student absenteeism, withdrawal, and the breakdown of the school-class system. These problems are probably common in many parts of the world. What are your thoughts on them?

Unger: As a medical scholar, I have a very simple approach to education. I think of my students as my children. Fathers and mothers want their children to experience the best things possible. Therefore, their primary role is to help children have experiences that nourish growth.

Ikeda: I find the principal spirit of education in your approach of seeing one's students as one's children. The best kinds of experiences are playing outdoors and bonding with other children. In modern industrialized nations, however, children spend far too much time indoors engrossed in computers and video games.

Unger: That is true. In today's information society many things, especially the mass media, have a tremendous impact on children. The daily recurrence of televised scenes of horrible violence clearly takes an educational toll. The appearance on the television screen of terrorist acts and serial murders is taken for granted.

Ikeda: Undeniably some high-quality television programmes cultivate and enrich children's sensitivities. But, by evoking hostility and anger, violently aggressive visual stimulation dulls imagination and empathy. Allowing children to confine themselves in environments where they do nothing but passively receive images weakens their abilities to think, judge, love and sympathize actively.

Family life and good reading, including the classics, can form a barrier protecting children from the corrupting influences of prevalent virtual reality. Reading greatly enriches the spiritual world of children. As an intellectual challenge it helps them pick and choose from the vast masses of media information and develop their own powers of active judgment and the imagination to empathize.

Unger: Children today experience a society of intolerance, war and aggression. In countering these influences, it is impossible to over-emphasize the importance of contacts with good books that cultivate sound judgment and imaginative powers. Strengthening the family is one way of reducing the bad influence of the mass media. Because in some ways the media are more influential than parents and family, my comment may seem unrealistic. But parents must serve as models and continue to put hope in education.

Ikeda: Children are the mirror of society, the era writ small. Today's problems with abnormal juvenile behaviours are rooted in the weakening of the educational influence that the family, community and school ought to exercise. In thinking of educational issues, we adults must see ourselves reflected in that mirror and be always on the lookout for ways to correct ourselves. As you say, good adult examples have a bearing on improving the power of education.

The Origin of Soka Education

Unger: What are the origins of Soka education?

Ikeda: Soka, or value-creating, education began with the humanistic educational approaches proposed by Tsunesaburo Makiguchi. Putting these into practice while serving as principal of an elementary school, he developed a system of value-creating pedagogy. He constantly insisted that children's happiness must be the goal of education. This was in the heyday of Japanese militarism, which mobilized every educational institution for the training of imperial and militaristic-nationalist youth. Nonetheless, he wanted to avoid children being sacrificed to the needs of society and to help every child live a happy life of limitlessly expanded potentialities. This wish is the basis of all aspects of value-creating pedagogy.

Makiguchi wrote:

> The important thing is the setting of a goal of well being and protection of all people, including oneself but not at the increase of self interest alone. In other words, the aim is the betterment of others and in doing so, one chooses ways that will yield personal benefit as well as benefit to others. It is a conscious effort to create a more harmonious community life.[20]

The fundamental aim of Soka education, then, could be described as to promote the happiness of both the self and the other and to foster individuals who are also able to do so.

Unger: I see. Makiguchi, founder of Soka Gakkai, advocated a value-creative education and demonstrated how faith can overcome the problems of life. His kind of humanistic education is essential to the cultivation of world citizens; that is, citizens who think and act on the global scale.

How does the worldwide Soka education network promote such education?

Ikeda: The Soka Junior and Senior High Schools, the starting point of today's Soka education network, opened in 1968. On that occasion, I set the following five precepts as goals for humanistic education:

1. Be a person of wisdom and passion, always pursuing truth and creating value.
2. Never cause problems for others, and be responsible for your actions.
3. Be kind and polite to others, rejecting violence, and value trust and harmony.
4. Express and courageously act on your convictions for the cause of justice.
5. Cherish an enterprising spirit, and grow to become honourable leaders in Japan and in the world.

Then, for the sake of the twenty-first century, I proposed five other principles:

1. Recognition of the unique dignity inherent in every life.
2. Respect for character.
3. Profound friendship enduring throughout life.
4. Rejection of violence.
5. The importance of the intellect and the need to be intellectual.

To my profound gratification, thanks to the unstinting efforts of all our faculty and staff members as well as the self-awareness of our students, these precepts and principles are being realized.

Unger: All of them are important guidelines.

Ikeda: From the time it was first opened, the students of Tokyo Soka Junior and Senior High Schools have sung their school song in which its lyrics inquire about purposes:

> *For what purpose do we refine our wisdom?*
> *For what purpose are we passionate?*
> *For what purpose do we love others?*
> *For what purpose do we strive for glory?*
> *For what purpose do we work for peace?*

Ignoring the profound goals embodied in these purposes can cause human beings and society to run wild. The tradition of the Soka schools entails deepening one's philosophy by constantly questioning one's purpose, creating a personal history through the actions taken during one's youth, and pioneering a path in life.

At the first entrance ceremony of the Kansai Soka Junior and Senior High Schools that originally began as girls' schools, I set another guideline: One should not build one's happiness on the unhappiness of others. I also said that, in comparison with the wide world, the Soka schools may be as small as a poppy seed, but if our students remain true to this ideal and practise the school guidelines, our impact will ultimately be felt all over the globe. This is because the underlying principle for lasting peace is one and universal. I said these things because I want our students to become strong and wise and contribute to the creation of happiness and peace for all wherever they go.

Unger: I can truly relate to that.

Ikeda: I ask the faculty members of the Soka schools and of Soka University to be first rate both academically and in character, and to be determined to improve themselves. I also expect that they create an educational institution where the students come first. From the students' viewpoint, the teachers themselves constitute the greatest educational environment. It is a fundamental of Soka education that, as leaders of humanistic education, teachers prize their students on a par with their own children. I want them to be the kind of teachers that students will be glad to have known, whose warmth they will appreciate, who will win their devotion, and to whom they will attribute their own achievements.

In 1970, shortly after its opening, Count Richard Coudenhove-Kalergi visited Soka Junior and Senior High School to talk with and encourage the pupils. He once wrote that knowing good and noble people was more useful in ennobling and developing people than everything else together.[21]

Unger: On the basis of my own experience, I know how important having wonderful teachers is to young students' humanistic education. A full, immersive interaction of teacher–child personalities with good teachers refines the balance among children with regards to their brains, bodies, and hearts.

Ikeda: Thomas Arnold (1795–1842), headmaster of the famous English public school Rugby, wrote that it was not a name, but the quality of the faculty that made a school excel or underperform. He also believed that teacher influence and mutual student-to-student influence fostered by teachers determine students' personalities.[22]

Humanistic education is character formation and refinement as a result of personal interactions between students and teachers who regard their charges with affection as great as their own parents. Because teachers define education, educational revolution must entail faculty revolution.

Unger: Your words help me understand why the eyes of students at

Soka schools and Soka University shine with hope. Because children are our most valuable treasures, we must choose teachers for them very carefully. Today, schools seem too negligent in this connection.

Personally, I would put maximum emphasis on schools, if I were in authority. We must train children to adopt a global viewpoint. Humanistic education enables them to perceive the whole world keenly. This means not only seeing the material, visible things between heaven and Earth, but also intuitively sensing spiritual values. Children who can understand things from a holistic perspective can learn to contribute to the wellbeing of all humanity. Humanistic education gives new meaning to the important ideas of ordinary natural-scientific education, thus making science more useful to humanity.

Ikeda: I concur. Without the wisdom to use it for human happiness, the most sophisticated knowledge is not only useless, but also potentially perilous. Toda used to say that the greatest delusion of modern humanity is to mistake knowledge for wisdom. On one level, knowledge can lead to weapons of mass destruction. It is undeniably true, however, that it can also lead to enormous convenience and industrial wealth. Humanistic education is in great demand as a way of guiding knowledge toward happiness and peace. In the years to come, the task of developing the wisdom to employ immense amounts of knowledge and information for the sake of human happiness through humanistic education is going to become increasingly essential.

In any event, the inner reformation within a single individual will undoubtedly inspire a transformation in those around them, unleashing this transformative force among common citizens to guide and shape public opinion worldwide. As this culminates into a groundswell of peace, blossoms of a new culture of peace will flower in rich profusion. The year 2007 commemorates the fiftieth anniversary of Toda's declaration against nuclear weapons. Those of us at SGI are determined to bring about a great tide of change in the times, from a culture of war to a culture of peace, sparing no effort to promote public education on disarmament and human rights.

CHAPTER FOUR

The Environment and Education

_____

Ikeda: The environmental issues confronting us demand urgent measures. We are constantly being warned of the destructive effect of global warming caused by carbon dioxide emissions from industry and other sources. Also, atmospheric pollution is depleting the ozone layer that protects Earth from hazardous cosmic rays. Imbalance between nature and humanity creates a crisis for the human race and the whole planet.

Unger: Our relationship to our Earth is shocking. We must understand the extent to which our planet is being exploited, how greatly we squander our natural resources, and how much we pollute the water and contaminate the air. We have only one Earth – only one environment to live in – and we all share concern about its possible destruction.

Ikeda: Very true. By the second half of the twentieth century, when the problem had become serious planet-wide, we finally grasped the magnitude of its scope. We realized that the natural resources we were squandering are not limitless after all.

Unger: Greater awareness of the global crisis can improve our sense of responsibility for the future. To save the planet means to save all life. In this sense, it represents the greatest justice, which we should not undermine. Fairness compels us to remember that, over the past fifteen years, a tremendous amount has been accomplished in Europe. I have in mind the halting of pollution, especially water pollution. Once gravely polluted bodies of water are now clean enough for human beings to bathe in.

Ikeda: A highly symbolic example. Recently, in Japan too, water quality has improved to the point that fish have returned to many formerly polluted rivers.

Unger: Equally symbolic is the increased attention being paid to animal welfare. But we still have one gigantic concern: poverty in the whole global environment. The only way to combat it is to cancel the debts of the poor and to cease exploiting them. This is, incidentally, an ancient biblical recommendation.

Ikeda: One of the most important problems in halting current global environmental destruction is dealing with the unplanned felling of forests in developing countries and the increasing exhaustion of arable lands. Conflicts of interest between developing and industrialized nations further complicate the situation.

At the 2002 Earth Summit in Johannesburg, developing nations questioned industrialized nations' right to promote a consumption culture while telling them to put up with poverty. In other words, a conflict arose over ways to balance economic growth and environmental conservation.

Unger: What annoys me terribly is the enormous gap between the affluent North and the impoverished South. Africa is a human tragedy to which the West has contributed. As the Johannesburg summit indicated, promises to do something about it remain empty as long as overexploitation proceeds unrestrained.

Ikeda: The only way to solve the problem of development–environment imbalance is for the industrialized nations to see beyond their own interests and adopt a global viewpoint.

Unger: I agree. But the hegemonic, authoritarian views the West has held prevent industrialized nations from taking a global view toward improving the environment and dealing with economic reconstruction of the developing nations. Some Western nations still desire to dominate everything left over from the colonial period.

Ikeda: Certainly as long as the industrial nations fail to change their way of operating, the global environmental problem will only become increasingly grave.

Unger: Yes. But I believe a fundamental paradigm change will take place in the twenty-first century to transform our values from lust for power and domination to willingness to act meaningfully. Acting meaningfully means overcoming attachment to one's immediate interests and materialistic desire and striving to build one's humane foundation through conscious efforts to discipline and develop oneself.

Ikeda: Instead of being controlled by desire, we must as individuals adopt loftier values and seek for a reformation from within and consummation of our own lives.

Unger: When industrialized nations embrace these values and address both environmental problems and the economic recovery of developing nations on a global scale, the global environment will improve significantly.

Ikeda: The key to a solution is a change in value criteria. Nichiren Buddhism teaches that treasures in a storehouse (economic affluence) are less important than treasures of the body (talents and social stature), and most importantly, the accumulation of the treasures of the heart (virtuous fortune) through altruistic action is most superior of all. Personal wealth and social position bring no true happiness. Our goal should be others' happiness as well as our own, which we can promote by refining and elevating our humane characteristics and overcoming the lust for power, status and material desires.

In *A Geography of Human Life*, published in 1903, Tsunesaburo Makiguchi was looking a century ahead when he proposed that humanity should advance from military, economic and political competition to an age of humanitarian competition. What he had in mind agrees closely with the change in value paradigm you mention.

Unger: The West lacks spiritual grounding and subscribes to hard-headed materialism. This reflects in the worsening environmental condition and an absence of planning for tomorrow. What is the Buddhist view on this situation?

Ikeda: In connection with relations between humanity and the environment, Nichiren Buddhism teaches that 'if the minds of living beings are impure, their land is also impure, but if their minds are pure, so is their land'.[1] To extrapolate, if human life is sullied with greed, antagonism and egoism, these evils will disturb the social environment and destroy the ecology. If, on the other hand, the human mind is filled with love, non-violence and reverence for nature, these good traits will harmonize human actions with the natural world and create a symbiotic social environment. Faith supplies the power enabling the good mind to triumph.

As you say, rejecting faith amounts to rejecting the transcendent and sacred. This undermines our power to control our internal evil, which leads to environmental destruction. Materialism that takes no thought of the future amounts to a rejection of the spirit and of ethics, which weakens the good and strengthens our inner iniquity.

Unger: In dealing with life, the physician or scientist must have reverence and respect, which grow with increasing knowledge of the relation between nature and life.

Ikeda: Reverence for life is the foundation and worldview of religion. Cultivating the feeling of reverence for life is religion's mission. Respect for the dignity of life must be the absolute and universal model for the twenty-first century because prizing anything else above life inevitably leads to oppression. Furthermore, reverence for life is the point of contact between science and religion, as I think you yourself have experienced.

Unger: I believe the age of reductionism is behind us. Anyone today who attempts to reduce all higher manifestations of life to simple, low-level processes becomes embroiled in contradictions.

Ikeda: The modern rational reductionist approach is insufficient in dealing with living phenomena, which are on a higher plane. Before the emergence of modern science, human beings derived essential nourishment from the natural environment, which they altered only to make living space for themselves. They maintained a sense of oneness with nature, for which they never lost their sense of awe. But, owing to the Industrial Revolution and advances in scientific technology, while deriving great material benefits, human beings allowed development to get out of hand and to affect the environment. Only when pollution began taking a heavy toll on people were we forced to recall that humankind is indeed inseparable from Mother Nature.

Since the time of Descartes, scientific progress has been thought to depend on such analytical methodologies as reducing difficult issues into simpler parts. In fact, however, all forms of life are organically connected in one great entity. The French-born American doctor and microbiologist René Dubos (1901–81), whom I met, stated 'Whether based on religious, philosophical, or social convictions, the feeling of significance derives from man's awareness, vague as it may be, that his whole being is related to the cosmos, to the past, to the future, and to the rest of mankind. Such a sense of universal relatedness is probably akin to religious experience.'[2] As you know, Western words for religion probably derive from Latin *religio*, to tie together. In other words, religion is a quest for connection between the individual and the sacred.

Each religion has a cosmology and worldview in which humanity has an assigned place and role. We can discover the answers to existential questions by recognizing our connections to the universe and to society – and that discovery would seem to be the source of our religiosity and our radiant humanity.

Unger: Investigating the meaning of one's existence is not a childish undertaking. Relying on religion is not clinging to delusion. Religion is a fundamental human desire to set life in the context of three great relationships: with nature, among ourselves, and with the spiritual. The challenge confronting us in the twenty-first century is finding the way for all humankind to cooperate for mutual growth and development. Our existence becomes stable only when the three elements of the triangle are in harmony. Humanity occupies the centre of the triangle.

Ikeda: What you say suggests to me the Buddhist principle of the 3,000 realms in a single thought moment of life expounded by Tiantai (538–97), a Buddhist scholar in China, and Nichiren. Explained briefly, this principle holds that the entire realm of phenomena, characterized as 3,000 realms, is contained in a single moment of life; that is, one instant of life. Each instant contains the potential of ten worlds[3] or states and is therefore pregnant with possibility. By the principle of the mutual possession of the ten worlds, each of the worlds contains within it the potential of all ten worlds, thus 100 worlds. Life at any moment manifests one of the ten worlds, or life-states.[4] Each of these worlds possesses the potential for all ten within itself, and this mutual inclusion of the ten worlds is represented as 100 possible worlds. Each of these 100 worlds possess ten factors, making 1,000 factors or potentials, and these operate within each of the three realms of existence,[5] thus making 3,000 realms. Your idea of the triangle of relations corresponds to the Buddhist doctrine of the Three Realms of Existence.

Because life at each moment includes the entire phenomenal world (the 3,000 realms), a transformation in one human thought or mind can transform not only oneself but also have a ripple effect throughout society and the environment. Thus, this principle is a philosophy of hope. Each moment of life contains universal wisdom and compassion. The goal of our movement, human revolution, is to evoke this wisdom and compassion and to act in solidarity to build a tranquil, peaceful society and environment.

Unger: I am aware of the Buddhist philosophy that strives to create peace in individual minds, in society and humanity, and in the natural world and its ecology.

Ikeda: In our dialogue, Dr Toynbee, demonstrating profound understanding of Buddhism, mentioned the importance and necessity of a world religion that enables humankind to realize that we are all members of a global community and all part of the whole cosmic life.[6] He expressed his hope that Mahayana philosophy will enable us to overcome greedy egoism.

Unger: You seem to interpret what, from the Christian standpoint, we call the Creator of the World as a spirit binding all things together. Christians consider love to be the bond between humanity and the Creator. The philosopher Gottfried Wilhelm Leibniz (1646–1716) emphasized this and spent his whole life working on the question of unity between divine and human reason.

Ikeda: Leibniz, as I understand it, insisted that a single mind is worth a whole world.[7] In other words, God is a mind, or spirit. The spirit of humanity was created by God in his own likeness and operates through God's true nature. On the basis of this strong connection between it and God, Leibniz believed that the value of the human spirit surpassed all other values.

Unger: In a certain way, you go still farther to see the spirit of God – what you would call the universal spirit – at work, not only in the human spirit, but also in all things. Christianity has not endowed the world of nature with such great value. As a result of the Fall, at the end of time Creation itself will need to be saved. Only then will God be all in all things. You do not recognize this kind of postponed deification. I have been using Christian terms like the 'spirit of God'. Are the same things applicable to Buddhism?

Ikeda: Buddhism teaches that the supremely precious Buddha nature is inherent equally in all living beings and can be revealed and manifested at any time and in all places. Insentient things, such as trees and land, in the world of nature too are endowed with the Buddha nature and also are capable of manifesting Buddhahood.

Buddhism also teaches that when each individual reveals their own Buddha nature, or the supremely precious value, from within, their distinctive characteristics can be allowed to reach full brilliance in their own ways just as the cherry, the plum, the peach, and the apricot all bloom in appealingly distinctive ways. The meaning of life is to manifest this supreme value fully, thus enriching the radiance of the self and the environment.

Another Buddhist doctrine called the oneness of life and its environment teaches that human beings and their environment are indivisible and essentially one. Our very lives thus determine the extent to which society and the natural world can manifest their values. In other words, Buddhism recognizes in human actions the power and responsibility to elevate and impart value to the whole world, including the realm of nature.

Unger: What you say provides much food for thought. To avoid misunderstanding, we must try to clarify the Buddhist and Christian understandings of *spirit*.

Ikeda: That is true. Another increasingly important factor in connection with environmental issues is examining how religions define and relate to nature. In February 2005 I spoke with Wangari Maathai, the first environmentalist to receive the Nobel Peace Prize. She was persecuted by her government for her democratic views, yet ultimately triumphed with her Green Belt Movement, which sponsored the planting of 30 million trees in the continent of Africa. She was deeply sympathetic with the Buddhist philosophy of attaching great importance to life, nature and the human community.

Unger: I see parallels between Christianity and Buddhism not only in terms of their deep concern and moderate attitude toward the Earth. They both take uncompromising approaches in healing and bringing salvation to man.

Ikeda: Speaking of environmental attitudes shared by Buddhism and Christianity calls to mind certain historical figures. One of them is

Saint Francis of Assisi (1181/2–1226), who, as you know, felt a close affinity with all creatures, whom he joyfully called his brothers and sisters. His great compassion for non-human creatures is illustrated by Giotto's famous picture of him preaching to birds. Believing that human life and the natural environment are indivisible, he cautioned that exploiting nature is a manifestation of greed. He had awe and compassionate affection for all creatures. In Saint Francis, I see something of the Buddhist bodhisattva.

Unger: That is an interesting comparison. The expulsion of humanity from the Garden of Eden should serve as a collective warning to us all: if, obsessed with the pleasures of the moment, we neglect to take sufficient care of the Earth, it will become uninhabitable. We therefore must treat nature with care and the sense of awe.

Ikeda: Precisely. If we control greed, have love towards humanity and regard nature with respect and if all society shares these values, we can create a symbiotic civilization that harmonizes with the Earth's ecology.

Unger: We must once again turn our eyes to the value of God's Creation and keep in mind the need for science to ensure a tolerable basis for life for all human beings.

Ikeda: Buddhism teaches how to live in symbiosis with and respect for nature. This way of life abides by the philosophy of the Middle Way, avoiding extremes of both asceticism and hedonism. In other words, while controlling the iniquity and base impulses inherent in life, the Buddhist view of nature and life promotes the building of an ethical foundation for a harmoniously symbiotic relationship between human beings and nature.

Unger: I agree completely with that philosophy. An excessively prosperous 20 per cent of the global population now consumes 80 per cent of the planet's natural resources. The remaining 80 per cent starves. In keeping with the Buddhist teachings, we must seek a way to control our own greed and live in communal prosperity with the peoples of the developing countries and with the natural environment. The Christians claim their mission is to go out in the world to learn while motivating themselves. This seems to me to approach the Buddhist ethos that prioritizes salvation of people above one's own enlightenment.

Ikeda: Buddhism conceives of bodhisattvas as active people concentrating on the salvation of others. Mahayana bodhisattvas take what are called the four universal vows. First, they vow to save all living beings; that is, to empathize with all suffering. Second, they vow to abandon all worldly passions. This means controlling all earthly desires and reforming them into such good attitudes as non-violence, compassion and hope. Third, they vow to learn all Buddhist teachings. In today's terms, this means learning Buddhism and the whole human spiritual heritage, including all fields of learning, philosophy and religion. Fourth and finally, they vow to attain enlightenment through Buddhist practice. In other terms, this means developing their own happiness as they save others.

These four universal vows well up from within the bodhisattvas themselves. This ethic is a pledge, spontaneous and self-directed, and not imposed externally. By proclaiming them, they manifest the goodness that directs them to their own and others' happiness and salvation. Surely the goodness itself generates an ethic making possible symbiosis with nature.

Unger: There is no sense in developing global or environmental ethics, unless they arise from the human heart. The desire to contribute to humanity and society has educational influence only when it is born from within us.

Ikeda: That is an important point. The strength born from within humanity is indispensable to orienting our values to ensure that science, economics, politics and all human endeavours are actually undertaken for the sake of humanity. Essentially, education, religion and philosophy should evoke inwardly generated human spirituality. I discussed the importance of this in a speech entitled 'The Age of Soft Power and Inner-motivated Philosophy', which I delivered at Harvard University in September 1991.

Unger: We must always have a volitional drive for education that has the power to move society. We discover the value of dialogue and education when we approach material issues of science and economics from the standpoint of the way human beings ought to live.

The individual human being is materially limited and capable of exerting little dramatic influence on the lives of others. Humanity as a whole, however, continues to live indefinitely in our progeny. It is therefore important that we globalize the value of individual service to the eternally continuing life of the whole human race.

Ikeda: I couldn't agree with you more. Shakyamuni stated, 'Whichever are seen or unseen, whichever live far or near, whether they already exist or are going to be, let all creatures be happy-minded'.[8]

This passage clearly elucidates an ethical standpoint toward all living beings, as well as an intergenerational ethic. Human beings are required to live in symbiosis with all of Earth's creatures; in other words, symbiosis with the 'seen or unseen' and 'far or near'. It is concurrently a stance of responsibility for the social and natural environment we inherited from those who lived before us, when we were born, and even more, a sense of responsibility to further improve and bequeath it to those yet to be born.

We must not jeopardize or deplete the rich and timeless flow of the human race. I also believe the key to solving the challenges to the global environment lies, as you explained, in 'globalizing the value of individual service to the eternally continuing life of the whole human race'.

First Step in Environmental Education: Grasp Extant Conditions

Ikeda: Earlier you mentioned that education has the power to move society. The starting point and driving force for a global and environmental ethic is education on the environment. How do you think environmental education should be promoted?

Unger: I would say it is impossible to separate it from other academic disciplines and concentrate solely on imparting knowledge. What we can do is to raise environmental awareness and train ourselves to take the environment into account in whatever we do. For instance, we have to continue reminding people of the extent of atmospheric pollution our actions bring, as well as our consumption of non-renewable resources.

Ikeda: The first step in environmental education is a firm grasp of present conditions: the amount of the world's forests already lost; the extent to which pollution of the atmosphere, the water, and the soil has already advanced; their ecological effects on the Earth, and so on. In connection with this point, in my proposal to the 2002 Earth Summit, I stressed the importance of comprehensively addressing the UN Decade of Education for Sustainable Development. As part of this undertaking, in cooperation with the Earth Charter Commission, SGI sponsors an exhibition entitled 'Seeds of Change: The Earth Charter and Human Potential'. It has already been shown in more than ten countries. Our '21st Century: Environmental Exhibition' also began touring Japan from 2006.

Unger: I am sure great things will come of the hard work the members of SGI are devoting to environmental education. As we stand at the threshold of the new millennium, we must use environmental education to promote symbiosis as a global model. One of the goals of incremental education is cultivating people who want to make the places where they live comfortable and safe for their neighbours as well as for themselves.

New Ethics of Control for Scientific Technology

Unger: The major problem in Europe is our highly secularized way of life, which generates enormous materialism enabling every person, as Nietzsche said, to be his or her own God. In other words, tremendous advances in the natural sciences have enabled us to exercise control over many things in ways that once would have been inconceivable.

Our present situation can be compared to that of Goethe's sorcerer's apprentice, who, enchanting a broomstick to fetch water but not knowing the magic word to stop it, is lost in a massive flood. The new era we are entering compels us to make completely novel ethical considerations to control science. We must learn to use the new instruments science puts at our disposal better than we have in the past, while rigorously advancing scientific research.

Ikeda: Insatiable intellectual curiosity and the spirit of inquiry have inspired research and technological innovation in many scientific areas. The pace of these changes is likely to accelerate in the years to come. Science and medical therapy have already entered realms once reserved to God, for instance in such things as cloning technology and advanced genetic engineering.

Now we are forced to draw a line between the technically feasible and the ethically permissible. How far should we be allowed to interfere in the manipulation of human life and nature? Where should we apply ethical brakes? These are some of the most pressing issues now confronting us. We will be unable to put our new scientific tools to use for the betterment of humanity unless we establish ethical models based on the philosophy of respect for the dignity of life.

Unger: Culture and science can survive only if they demonstrate respect for life and the environment. As is evident from many psychological sicknesses, exploiting nature means exploiting ourselves too, thus generating psychosomatic illnesses.

Ikeda: I have discussed non-violence toward the environment with the Indian agronomist Dr M.S. Swaminathan, who, referring to Gandhi on this topic, said human beings will use violence towards nature as long as they use it among themselves. Teaching non-violence is the most important aspect of environmental education, which itself must be founded on respect for the dignity of life. Education must persist in underscoring the potential of each irreplaceable life and the dignity of all life that is their mainstay.

Unger: I agree. Culture is born, not of exploitation, but of respect. Education can be environmental in nature only if it teaches respect for life.

When we concentrate on our own environment alone, we lose our point of reference with the transcendental. And without that we cannot conduct environmental education in the fullest sense.

Ikeda: Children, who bear the responsibility for the future, must be taught about respect for the dignity of life and reverence for the transcendental as part of environmental education. In *A Geography of Human Life*, Makiguchi noted two things to remember in relation to the environment. First, we can intellectually recognize the laws and order of nature, but we should not forget that the laws and order of nature are not the products of human intelligence. Second, on the emotional plane, we must be aware of the need for a sense of piety and awe toward the religious sphere in which these natural laws are founded.

An intellectual approach alone can lead to the arrogant idea that science is omniscient. A solely emotional approach can lead to disassociation from the realities of life. Makiguchi argued that the correct attitude toward nature combines the intellectual elements of rule and order with the emotional element of awe toward the religious or transcendental.

Unger: Such an attitude toward the natural environment is important to enabling children to grow up with mind, body and spirit in a state of balance.

Ikeda: When you visited the Kansai Soka Junior and Senior High Schools in July 1997, you said that work must be done with the head, the heart and the hands. You remarked that work done with the head and the intellect alone, could be cold. Done with the emotions alone, it loses touch with reality. But done without the intellect, without the emotions, and with the hands alone, it could lead to the destruction of the planet. That is why we must live in a way in which all three are balanced.

In these clear terms, you pinpointed the origin of the present global crisis and humanity's loss of harmony. I am grateful for the way you taught our young people the importance of growing up as whole human beings in which cool minds, warm hearts and speedy action are harmonized.

Unger: No philosophy or way of thought has meaning if it fails to convey itself to others. Because I wanted to explain my meaning to young people responsible for the new century, I spoke simply. I was greatly impressed by the animation of the students I have met on my visits to the Kansai Soka schools and Soka University in Tokyo.

I wanted to help them become – to borrow your words – whole, well-balanced beings.

Ikeda: Makiguchi thought that the aim of education is not to cram students' heads with fragmentary information but to develop their whole beings to enable them to use the knowledge they absorb for the sake of humanity. He proposed what he called the half-day school system for the sake of cultivating all-round personalities capable of using head, heart and actions. According to his system, students did classroom work for one half of the school day and practical work during the other half. The cultivation of the all-round personality is one reason why Soka University of America was founded as a liberal arts college.

Unger: We expect the Soka schools and Soka University to turn out large numbers of well-rounded human beings qualified to improve the environment and create a brighter future.

Nuclear Weapons as Life's Evil Aspect

Ikeda: It is imperative to develop wisdom, through which we come to the understanding of for what purpose we use knowledge – and environmental issues and nuclear weapons are representative of this point. Nuclear weapons are the product of scientific knowledge. In considering how human beings should deal with them, the issue of knowledge and wisdom becomes acute. The end of the Cold War provided an excellent opportunity for getting rid of nuclear weapons entirely; and indeed, during the final decade of the twentieth century, the movement to abolish them made some headway and moved beyond nuclear arms reduction toward the creation of an international order in which their use would be banned.

For instance, the International Court of Justice issued the advisory opinion that the use of nuclear weapons violates international law.

In 2000, in its final document, a review conference for the Nuclear Non-Proliferation Treaty (NPT) called for a clear promise to eliminate nuclear weapons completely. But then the United States' concept of missile defence led to an upset of the nuclear balance and the start of competitive nuclear expansion into outer space. Furthermore, both the United States and Russia conduct subcritical nuclear experiments not specifically forbidden by the Comprehensive Nuclear Test Ban Treaty (CTBT). How do you think humanity should confront the still persisting threat of nuclear weapons?

Unger: I am truly no friend of weaponry. It frightens me. Today humanity already has sufficient weapons to wipe out the whole world. As it develops globally, though hypocritically describing itself as peace-oriented, the weapons industry actually becomes increasingly deeply involved in war. I have learned, however, that nations need weapons to defend themselves. Still I am very disturbed by nuclear, biological and chemical weapons. Using them or nuclear weapons is the most cowardly thing human beings can do. My personal opinion is that Hiroshima and Nagasaki should have taught us that using nuclear weapons is wrong.

Ikeda: Given that nuclear weapons are capable of annihilating the human race and destroying the ecological system, the question, like the environmental question, boils down to whether we prefer the interests of nations above the interests of all humanity and the whole planet. There can be no mistake: the key issue is between humanity and nuclear weapons. In this connection, I must once again refer to Toda's declaration against nuclear weapons, in which he said: 'Even if a country should conquer the world through the use of nuclear weapons, the conquerors must be viewed as devils, as evil incarnate.' This statement breaks through political and military barriers to reach the fundamental dimension of the dignity of life itself. Buddhism teaches that inherent in life is the evil impulse striving to fulfil selfish desires even at the cost of destroying the lives of others and even the natural environment. The state in which one is controlled by this evil is termed the Realm of Freely Enjoying Things Conjured by Others. For nations, ethnic groups, or people in authority, to attempt to dominate others by means of nuclear weapons represents such evil in its purest form. It is the evil that Toda asserted we must crush completely.

Unger: I agree with him. I have a simple answer to the nuclear question: nuclear weapons must be eliminated.

Ikeda: We must learn to assign first priority in our actions to the advantage of humanity and the planet. The famous Russian cosmonaut Dr Alexandr Serebrov, with whom I have published a dialogue entitled *The Cosmos, the Earth, and Humanity*, told me that viewing the Earth from space not only provides a valuable and philosophically pious experience, but also stimulates a sense of mission to do something for our precious planet. The American astronaut Dr Donald K. Slayton (1924–93) once told me that his similar experience taught him that the cosmos provides a unifying context for the Earth and is the locus of human unity. Today we must put the interests of humanity and the Earth first. To do this, we need to adopt the viewpoint of these men who have observed the planet from space.

What hopes do you have for the space age in the twenty-first century?

Unger: When we draw up rules for its use, we must consider that space belongs to everyone and is the fabric of our existence. As our only possible habitat, it demands our eternal cooperation. Indeed, it is the starting point of human cooperation. Consequently, at the beginning of this new age we must not misuse space development.

Ikeda: Individual nations must not be allowed to monopolize space development in their own interests. Competitive arms races must be strictly outlawed. In all major religions, prayer amounts to dialogue and resonance with the eternal universe, the progenitor of both human and non-human life. In this sense, as you say, the cosmos is the fabric of our existence. Have our hearts and minds grown as expansively as globalization has? Today, the powerful influences of narrow nationalism and national egos make the role of spiritual exchanges with the cosmos immeasurably important.

Unger: I agree. It is especially vital for young people, who are responsible for the future, to have opportunities for cosmic spiritual exchanges and to develop a universal viewpoint. I believe the Kansai Soka schools participate in an educational programme inspiring a sense of intimacy with the cosmos.

Ikeda: Yes. They participate in Earth Knowledge Acquired by Middle School Students (EarthKam). A programme held under the auspices of NASA (National Aeronautics and Space Administration) in the US, EarthKam makes it possible to take photographic images of any part of the Earth's surface using remotely controlled digital cameras mounted on the International Space Station. Students calculate the space station's orbit and the locus and time of the photograph and then transmit photographic instructions in English. These instructions are transmitted to the space station via an organization like NASA, and the photograph is taken. This system enables students to perceive visually aspects of the face of the planet that are difficult to grasp in ordinary class work. The Kansai Soka Junior and Senior High Schools hold the world record for the number of times participating in EarthKam. The Tokyo Soka Junior and Senior High Schools, too, are conducting space education in cooperation with a world-famous observatory. These opportunities enable students to see the world as a small planet in space totally without national boundaries and in need of our diligent care.

Unger: These wonderful opportunities to see the Earth from space extend your kind of environmental education outward into the cosmos.

Ikeda: For the coming era, we must rework the famous admonition from 'think globally, act locally' to 'think cosmically, act globally'. The human race must realize that we children of Earth are all fated to live together on our planet. Our times require humanistic education that teaches us to adopt a universal approach and to act as citizens of the cosmos in our attempts to deal with the global issues confronting us.

CHAPTER FIVE

Health, Medicine and Bioethics

Ikeda: Making the current century a century of life has been our con-sistent commitment. We all hope the twenty-first will be a century in which good health, illness, mortality, and respect for life will be focal points of increasing attention. In this connection, in this final part of our discussion, I would like to discuss life and bioethics.

Unger: In light of rapid developments in bioscience, I maintain that it is vital we look more deeply into life itself. I am delighted to have this opportunity to explore a philosophy of life through this dialogue.

Ikeda: I feel the same way. You are now a world authority in cardiac surgery. Why did you select this field?

Unger: I always wanted to be a doctor. When I was a medical student, cardiac surgery was still difficult and dangerous, and I decided to do something to improve the situation.

Ikeda: Having chosen this difficult path, you have saved many precious lives. I understand you have performed surgery on more than 7,000 patients.

Unger: Yes, about that many.

Ikeda: How, in your opinion, can we remain in good health?

Unger: Two conditions determine good health. First is freedom from the possible presence of genetic abnormalities. Second is the effort the individual makes to remain healthy. This means observing dangers to good health and dealing with them. The person who cannot do this runs the risk of sickness. Beyond the age of forty, each person must be his or her own doctor.

Ikeda: Very useful and valuable advice. In other words, we require the wisdom and resourcefulness to protect our good health by being our own doctors and nurses. What is your opinion of the currently held idea that a stressful society causes sickness?

Unger: Stress has two aspects: a positive and a negative. Challenges stimulate positive stress. By overcoming a challenge, human beings develop psychologically and physically. Without it we would grow lazy. Negative stress, on the other hand, is dangerous. When dominated by it, people become dispirited and pessimistic, losing even the will to accept challenges and ultimately falling ill.

Ikeda: I see. Whether stress operates positively or negatively is determined by the way we confront it. That is why it is important to bravely confront challenges with a sense of purpose in life. We must find ways to ignite the life force inherent in us. I believe associating with people in whom life force is abundant and being in environments rich in those forces is important.

Maximum Human Lifespan

Ikeda: What kinds of people do you think are able to lead long lives?

Unger: Of course, none of us can live forever. But chances are good for greater longevity in people who have already lived beyond the age of sixty.

Ikeda: In the East, the age of sixty has long been afforded special significance as a major stage in the human life cycle. My own observations of many people suggest that sixty is in fact a milestone.

Unger: Some people are genetically conditioned to live long. Beyond genetics, however, physical and spiritual exercise is necessary to longevity. Indeed, the latter is the key.

Ikeda: I assume you mean that activity on the mental level, like faith and conviction, influences long life.

Unger: Certainly. Faith and conviction are the opposite of fear. Having a foundation of faith means knowing no fear. And this is important to tolerance. Faith enables people to make correct judgments and gives them stability.

Ikeda: Firm faith and conviction put us on the track to good health and long life. In medical terms, how long can life be extended?

Unger: Unless there are obstacles, I think a person can live 120 years maximally. The end of life is determined by our genes.

Ikeda: Really! By a wonderful coincidence, the same age is given as a possibility in a Buddhist scripture, Tiantai's *Words and Phrases of the Golden Light Sutra*.

Unger: Human cells are said to be renewable in one form or another until the age of 120; thereafter they lose the ability.

Ikeda: Buddhist teachings and modern medical science agree on human longevity in a remarkable way. Although human beings are naturally greatly interested in how long they will live, actually quality of life is far more important than its length. In *The Dhammapada* we find this passage: 'Better to live in strength and wisdom for one day than to lead a weak and idle life for one hundred years.'[1] In one of his writings, 'The Three Kinds of Treasure', Nichiren wrote, 'But it is better to live a single day with honor than to live to 120 and die in disgrace'.[2]

With your great experience in cardiac surgery, are you able to identify hearts that promise longevity?

Unger: Each human heart is different. During surgery, examination of contraction and colour may lead us to assume that, after successful surgery, an operated heart may live at least another twenty years.

Ikeda: In Japan, where average life expectancy is longer than anywhere else in the world, cardiac illnesses come second after cancer as a cause of death. What can we do in daily life to minimize the danger of heart disease?

Unger: Nutrition is a primary concern. In rural Japanese villages, cancer – especially of the oesophagus – causes many deaths, whereas cases of cardiac illness are comparatively few. In Tokyo – as in other big cities like Los Angeles and San Francisco where people eat rich diets heavy in meat, deaths from heart disease are numerous.

Ikeda: Until about the 1970s, the leading cause of death in Japanese rural villages was cerebrovascular disease followed by cancer, and cases of cardiac illness were relatively rare. At present, however, as rural people, too, eat more meat, cardiac illness is on the increase. In his *Great Concentration and Insight*, Tiantai listed disorderly dietary habits as a cause of illness. Another Buddhist scripture, Gocaropaya Sutra, lists the following effects of overeating:

1. Laziness,
2. oversleeping that causes other people trouble and
3. upset physical condition and sickness.

Unger: Stated simply, the way to prevent cardiac sickness is to follow these three rules:

1. Don't smoke,
2. reduce intake of foods high in cholesterol and
3. in diabetics, scrupulous monitoring of blood sugar.

Ikeda: Very clearly set forth! I have heard that a drink of water before sleep prevents dehydration, promotes good circulation of the blood, and contributes to good health. Is that so?

Unger: Improving the circulation protects the brain. Getting plenty of water is important. It is said that people free of heart or kidney problems should drink about two litres a day. We should be cautious not to drink too much mineral water because it is high in sodium, which contributes to hypertension.

Ikeda: What foods do you recommend as good for the heart?

Unger: The best thing is to eat only half of everything. Cut back on meat to only about once a week. But eat fish, fruits, grains and vegetables. A glass of good wine is fine too. But balance is all-important. Overeating and over-drinking inevitably lead to sickness.

Ikeda: Moderation in diet is the key. *The Dharma Analysis Treasury* explains four categories of nourishment:

1. Foods actually put in the mouth like meat, fish, and vegetables,
2. things that give joy and pleasure on contact like good music and fine art,
3. things that invigorate by inspiring thought and giving hope, and
4. mental power inspiring the will to live.

All that which provides energy for living, not just food, is considered nourishment.

The mutual interrelation of these four sources of energy maintains good health. What kind of exercise is best for the heart?

Unger: Walking every day. Climbing stairs, too, is good. I generally do not ride in elevators. I believe climbing three flights of stairs is better than jogging for an hour or a workout at a gymnasium.

Ikeda: It certainly is wisdom to maintain good health. A Buddhist walking exercise that consists of walking back and forth in a fixed space regulates the physical condition. *The Fourfold Rules of Discipline* attributes five effects to it:

1. Increases the ability to walk far,
2. promotes reflection,
3. reduces the likelihood of illness,
4. promotes digestion and
5. prolongs the steadiness of mind.

Gandhi wrote that 'no matter what amount of work one has, one should always find some time for exercise, just as one does for one's meals. It is my humble opinion that, far from taking away from one's capacity for work, it adds to it.'[3]

What is the best time of day for exercise?

Unger: Morning, because it stimulates circulation. A little morning exercise raises the pulse. This makes it easier to deal with the various stresses encountered throughout the day.

Ikeda: What ways of combating stress are good for the heart?

Unger: Again, walking a little. Getting the right amount of sleep, too, is important, as are loving others and being loved by others.

Ikeda: A simple statement of important truths. Although individuals differ, what is the ideal amount of sleep time per day?

Unger: It varies with age and physical needs. Older people require less. Active people in the prime of life require more. On the other hand, the young can go without sleep for a little while. A short nap after lunch is effective. This is why Mediterranean peoples customarily take a siesta.

Other adjustments should be made to the activities of the day. For instance, in hot weather it is better to work during the cool morning hours and rest in the heat of the day. When adjusting the temperature is impossible, we can make adjustments in our schedules. It is important not to rush but to work as steadily as possible.

Ikeda: Your words about good health reflect sound wisdom.

New Light on Origins of Life and Evolution

Ikeda: All terrestrial life forms are composed of proteins genetically determined through DNA. They are all descended from the first DNA on Earth, which over a period of four billion years has ramified into myriad forms, some of which are now extinct. This is the way modern medical science and molecular biology explain the origins of life and the mechanism of evolution. How do you evaluate it?

Unger: In the course of human development, we have come up with many biological evolutionary models. In the twentieth century, molecular biology taught us that genetic information is inscribed as a DNA base sequence. The discovery of DNA enabled us to explain biological evolution on the basis of genetic mutation. We know that all life forms on Earth descend from the original DNA and, at the molecular level, are made in the same way.

Developments in today's genetic engineering provide the basis for explaining biological evolution genetically. Traditional biology could not explain the origin and evolution of life. Genetic engineering gives us new knowledge for doing so. No doubt further genetic engineering developments will result in still more detailed elucidation. Genetic technology will take us to a totally different dimension and exert grave influence on our lives. I believe it will help us better understand God's creation.

Ikeda: Symbolic of the results of genetic engineering, the decoding of the human genome in 2003 showed that human beings have some 30,000 genes.

Unger: Yes. There can be no doubt that month-by-month advances in genetic engineering are having an increasing impact on biology. In the future, genome decoding will enable us to learn even more about genes, thus permitting us to develop diagnoses and therapies for genetically related illnesses. We will be able to prevent or arrest illnesses. Genetic engineering will be indispensable in the pharmaceutical industry in the development of many products, such as the insulin required by diabetics.

Ikeda: Undeniably humanity can look forward to many advantages from genetic engineering. Genetic diagnostics and therapy are sure to become increasingly common.

Unger: Certainly new kinds of diagnostics and therapy will be developed.

Ikeda: But there are people who consider rapid advances in this area dangerous manipulation of life itself. Because of its central importance to human existence we must cautiously consider the negative aspects of genetic research. As some people point out, the discovery of a gene predisposing a person to a certain illness might prejudice their chances for employment or acceptance in an insurance plan.

Unger: Not all genetic information has been completely decoded. As you point out, decoding of the genetic information can become, not enhancing, but threatening. For instance, conceivably a person with a gene making cancer by the age of sixty a likelihood might be refused insurance or rejected by a prospective employer. At our present stage of knowledge, such a diagnosis is premature and mistaken.

Ikeda: I agree. Using genetic information to invade privacy and as a cause of rejection for insurance, employment or marriage is putting the cart before the horse. A blueprint of a life, genetic information is the ultimate area of human privacy. That is why protecting it and using it effectively must always be taken into consideration in genetic diagnostics. Science must protect human dignity.

Unger: The European Academy of Sciences and Arts, Salzburg, organized a research conference on this topic with the National Academy of Medicine in Washington DC in 2001. The topic of this conference, 'The Impact of Gene Technology on the Human Dimension', was treated in two segments. In the first, specialists discussed the most recent results in gene technology. In the second, philosophers, theologians and economists examined the human dimension.

Ikeda: For genetic technology to serve the cause of human happiness it must take into consideration the opinions of – as you say – philosophers, theologians, economists and others.

Unger: Genetic technology affects all fields. Our conference highlighted grave negative factors as well as high expectations and hopes. One of the hopes is that genetic technology will give birth to a new industry to recompense the industrial world for the considerable investments it made in the genome project.

Ikeda: The focal point is how genetic technology can be linked to the enhancement of human happiness and the negative aspects of this technology kept under control. What is your position on this?

Unger: I represent the natural-scientific viewpoint that human beings should study everything that exists. Science provides us with plenty of information helpful in overcoming diseases and enriching human life. Genetic engineering has the power to revolutionize our way of thinking and benefiting humanity.

Reproductive Therapy and the Use of Embryos

Ikeda: Today medical science goes beyond therapy and prevention to manipulate human life itself. The first human ovum that was fertilized in-vitro occurred in 1978. Since then reproductive therapy has advanced with astonishing speed.

Unger: Yes, it has. Originally developed for veterinarian use, in-vitro fertilization is now fulfilling the hopes of many couples who, though wanting them, have been unable to have children. In-vitro fertilization itself is an assistance technique whereby a process that normally takes place within a woman's body is conducted outside it.

Ikeda: As you say, it has been a great blessing to many apparently infertile women. But it has also opened the door to therapeutic intervention in the early stage of life. It is now possible to select fertilized ova and manipulate cryogenically preserved embryos. Doctors and scientists engaged in this field should be guided by respect for the dignity of life and focused on the objective of enhancing human happiness in the form of overcoming infertility.

Unger: I, too, am opposed to limitless intervention in the origins of life. On the other hand, a bigger problem today entails fertilization, not for reproduction, but for regenerative therapies in other medical fields. For instance, 'stem cell' technology has made possible the production of various other kinds of cells from the fertilized ovum at the blastocyst stage or from the cells of an aborted embryo.

Ikeda: This field is promising in connection with the treatment of intractable pathologies and in regenerative medicine. But the inevitable destruction of embryos entails ethical problems. The Japanese Council for Science and Technology Policy is currently debating the issue of embryo use. Unfortunately, however, they adopt as their criterion the usefulness of embryonic research instead of the proper treatment of life itself.

Unger: But we must remember that many embryos fail to reach the stage at which they can be called life. Not all artificially inseminated ova are returned to the mothers' bodies. A certain number of them must be discarded. Even many naturally inseminated ova are never implanted and are wasted. I believe human life starts when the embryo is settled in the mother's uterus. Therefore, I see no great problem in using as research material inseminated ova that are not returned to the mothers' bodies. Still we must remember what the late Cardinal Franz König, who was honorary chairman of the European Academy of Sciences and Arts, said: Researchers must do research, but they must also abide by the voice of conscience.

Ikeda: Although it can be a great blessing, stem cell research can also threaten the dignity of life. We must, therefore, be very wary. Tampering with the embryo, the bud of life, runs the danger of regarding human beings as a means to an end, which relates directly to the loss of human dignity. As it proceeds, embryo research must always be cognisant of the nature of the embryo as the bud of human life.

Brain Death and Organ Transplants

Ikeda: Brain death occurs when the brain has irremediably ceased functioning and the heart is kept beating on artificial support systems. In Japan, opinion is divided as to whether this state constitutes death.

Unger: I hesitate to use the term 'brain death' because people tend to think that a person in this condition is in fact already dead. If such is the case, discontinuing intensive care becomes permissible. I prefer the French neurological term *coma dépassé* (deep coma). In this state, the brain is disordered, the body temperature drops, the heart still functions but circulation drops, and the nails turn bluish-black. Somehow or other *coma dépassé* was mistranslated into English as 'brain death', and the term has stuck.

Ikeda: The focal point of this issue lies in relation to organ transplants, because they often depend on people in the so-called brain dead condition, which artificial support systems can extend for long periods. What are your thoughts on the topic?

Unger: Brain death is determined with a purpose in mind – organ transplants for instance. In such cases, a decision needs to be made whether to discontinue intensive care. This determination serves as justification for removing organs from donors. Some people fear that organs may be removed from still-living subjects. Actually, however, international practice has established certain relatively objective standards for judging brain death. These must be strictly adhered to.

Ikeda: Some time ago, I proposed stringent criteria which, with the addition of the condition of absence of circulation in the brain, should alleviate the anxiety suffered by patients' families. Because of the acute scarcity of donors in Japan, a movement is underway to ease conditions for transplants from those in a state of *coma dépassé*. Until now, transplants have been authorized on the strength of both signed donor cards and family consent. The revised system would authorize them on the basis of family consent only.

I believe, however, that the spontaneous, un-coerced wish of the patient should be the premise of brain-death organ transplants.

Different people define death differently. Authorizing organ donations solely based on family consent might mean that organs are removed from the bodies of people who do not equate brain death with death. Furthermore, in many instances today, families afflicted by having a member sick enough to become brain dead find it difficult to give their full consent on the basis of doctors' explanations which, owing to time restrictions, may be unsatisfyingly brief. When it is impossible to know the patient's attitude toward brain death, death and organ donation, serious consideration must be made of the great burden the revised standards put on surviving family members.

Unger: Some people are afraid that organ transplants from the brain dead reduce the patient from a subject deserving therapy to an object from which organs may be harvested. Personally, however, I believe that using the term *coma dépassé* preserves the dignity of a donor who has already clearly expressed willingness to give another subject vital organs. This is an act of life. But the practice of selling organs for transplants – which is said to occur in some countries – must be strictly rejected as it seriously damages trust in medical treatment.

Ikeda: I agree completely. To compensate for the scarcity of human organs, transplanting organs from non-human animals is under consideration. For instance, pig-heart transplantation is being discussed. What is your response?

Unger: This is a very complicated issue. I react negatively because of the high likelihood of transplanting the animal's viruses along with the organ.

Ikeda: We must then put our hopes in artificial organs. In 1986, you began developing the smallest artificial heart and have become world famous as a pioneer in the field. The project must have been arduous.

Unger: I started developing an artificial heart (ellipsoid heart) in 1975 and implanted it in 1986 clinically as a bridge toward transplantation. This has been a pioneering step and twenty years later this step turned out to be a large one. The artificial heart must be very simple; otherwise it fails to function as it should. Subsequent technological innovations led to the development of an artificial heart for clinical use. The future looks bright. The artificial heart will someday become as easy to use as the pacemaker is today.

Euthanasia

Ikeda: The legalization of euthanasia in Holland in 2001 and in Belgium in 2002 has cast death-related bioethics in sharp relief. The salient point is determining whether it is permissible deliberately to shorten life in order to terminate the suffering of an incurable patient. What are your opinions?

Unger: Personally I oppose euthanasia. It is impermissible from both the medical and the humane standpoints. The doctor's job is to use therapy to prolong, not to destroy life. The debate about euthanasia did not arise in the clinic really. It is a discussion of whether to recognize the value of the lives of people with protracted consciousness disorder (the so-called vegetative state) or of those with grave intellectual disorders. In other words, it is a discussion of whether it is permissible to curtail lives that are not worth living. This was the theoretical basis of the Nazi euthanasia plan. It is not the doctor's job to judge the quality of life. This is why I feel it is mistaken for doctors to talk glibly about euthanasia. They should discuss how to free terminally ill patients from pain, not ways of killing them.

Ikeda: I agree with you. Doctors should make every effort to relieve pain through the use of anaesthesia and analgesic therapy. Therapy that overextends life is thought by some to worsen the patient's suffering and diminish his or her dignity. This was one of the ideas behind the recent legalization of euthanasia.

Actually, however, at present considerations of human dignity miss the mark. They tend to concentrate on cognition, or on the personality as a function of cognition. Consequently, cognitive regression or irremediable loss of the psychosomatic functions is thought to detract from human dignity. On the basis of this approach, euthanasia may be forced on weak patients with intractable illnesses. Such a situation should not be allowed to occur.

Near-death Experience

Ikeda: On the many occasions when you have witnessed patients confront death have you had any experiences indicative of the wondrous nature of life and the greatness of the strength to live?

Unger: On several occasions I have observed in awe while life has left the tissues when a patient has died during surgery. One senses the soul leaving the body.

Ikeda: Scientific reports compiled in the United States on near-death experiences speak of seeing rings of light, of being drawn into a dark tunnel, and of departing from the physical body and observing oneself during surgery. Have any of your patients had similar near-death experiences? What are your ideas on them?

Unger: I have known many patients who after several days in a coma state have come back to life after having been pronounced unlikely to regain consciousness. All of them have said that, during the coma, they were bathed in a wonderful warm light as they travelled a long, narrow tube and that they were reluctant to return to this world. I think these near-death experiences reveal to us the meaning of our existence. Perhaps being bathed in light we profoundly long to approach is the meaning of existence.

Ikeda: The Swiss-American psychiatrist Elisabeth Kübler-Ross (1926–2004) describes her own experience of having left her body and of having stood over the bed looking down on herself. She also speaks of approaching and melding with a light. Her description has much in common with the near-death experiences you mention.

The mention of a light brings to mind Tolstoy's story *The Death of Ivan Ilyich*, in which an ordinary government official falls fatally ill and struggles with the fear of dying until ultimately he reaches the pure happiness of eternal light. Transcending time, light symbolizes the leap to eternal life.

The Buddhist teaching of the nine consciousnesses deals with deep levels of conscious awareness and indicates that after death, the individual life becomes infused with the eternal, fundamental life of the universe; that is, the cosmic life.

Near-death experiences, shared by peoples of totally different backgrounds, include common, universal elements transcending culture and religion. Though much about them remains unexplained, they are nonetheless fascinating.

Eternal Nature of Life

Ikeda: Together with the question of death we confront the difficult issue of the eternal nature of life. Buddhism expounds that life is eternal, continually repeating the cycle of birth and death. What are your ideas about the eternal nature of life?

Unger: This is a very difficult issue. Christianity teaches that the soul with the spirit is in the body until death, when it is liberated to rise and sublimate with eternal life.

Gene technological developments do not invalidate this way of thinking. Let us assume that an abnormality for blindness occurs in the chromosomes of a fly. The offspring of that fly, too, will be blind. The same kind of thing happens in all animals and plants, including human beings. Our religion teaches that the Creator, God, gives us life. But if life is determined by genes, or even by chains of proteins, how are we to know where we came from and where we are headed? Once again we encounter the old fundamental problem. Perhaps we can only sense the eternity of life with both our conscious and our subconscious. Perhaps the only way to find clues about eternal life is to refine our instinctive intuition.

What does the doctrine of the nine consciousnesses teach about eternal life?

Ikeda: The first five of the nine consist of sensory perceptions – sight, hearing, smell, taste and touch – received through the eyes, ears, nose, tongue and body. The sixth consciousness compiles, analyses, and makes judgment on information received through the five senses. It is primarily with these six functions of life that we perform our daily activities. On still deeper levels are the seventh, or *mana*, consciousness (inner spiritual world independent of sensory input) and the eighth, or *alaya* (karmic storehouse), consciousness. At this point we are dealing with depth consciousness or the subconscious.

Unger: The interplay of the conscious and the subconscious has become an important field of research. Recent cerebral science has clarified cranial-nerve activities that precede consciousness and free will. Measurements of cerebral structures make it necessary to suppose a subconscious intent behind conscious acts. I understand that our living in this world arises from the subconscious, expanding widely into the conscious.

Ikeda: That is a very stimulating idea. In modern terms, the karma that according to the teachings of Buddhism continues through life–death cycles could be called potential life energy. The *alaya* consciousness pertains to the continuity of karma. At the death of the individual, the operations of all seven other consciousnesses become latent in this eighth consciousness, which is said to continue to operate. When we discussed this issue, the world famous physicist and former rector of Moscow State University Anatoli A. Logunov (1926–2015) expressed great interest in the enduring nature of karma.

On a level still deeper than the *alaya* consciousness, Buddhism postulates a ninth, *amala* (fundamental, pure) consciousness, which can be thought of as universal life itself. You mentioned that our living in this world arises from the subconscious, expanding widely into the conscious. Death makes the individual life latent in cosmic life. Birth represents manifestation in the *mana* and other consciousnesses from the *alaya* consciousness.

Unger: Sensing eternal life probably requires the application of both the conscious and the subconscious and refinement of our instinctive intuition.

Medical Ethics and Medical Practice

Ikeda: All doctors swear to abide by the Hippocratic Oath, which binds them to apply all required measures for the benefit of the sick, keep them from harm and injustice, give no deadly drugs, use harmless principles, guard patients' confidentiality, give no abortion-inducing medicines, and so on.

Unger: The Hippocratic Oath is a universal ethic for doctors because it teaches us our duties to relieve pain, heal, and cure illness.

Ikeda: What is your concept of the ideal physician?

Unger: I always say we need doctors, not medical technicians. By *doctor* I mean a person whose personality is radiant in totality. A person who is warmly humane. A person who has a balanced sense enabling them to discriminate between the important and the trivial. Such a doctor must keep in mind the usual ethical admonitions to honour parents, not to kill or steal, and so on. As medical science progresses, doctors are required to possess various kinds of knowledge and technical skills. Therapeutic facilities grow increasingly complex. This is why fundamental ethics of life are so important.

Ikeda: Very clear. What conditions must doctors fulfil?

Unger: Day and night, doctors must protect and work for their patients.

Ikeda: You state it concisely and with a ring of philosophical conviction. The Buddhist scripture called 'Sovereign Kings of the Golden Light Sutra' teaches that in all cases doctors must be compassionate and free from lust for profit. Doctors exist to relieve, never to profit from, patients' suffering. They must therefore always have a compassionate heart. You embody the meaning of working for the sake of patients.

Unger: You are kind to say so. Striving to be that way has become a part of my life.

Ikeda: I imagine that you approach each surgical operation as a battlefield where the struggle for victory is constant. Are there points you keep in mind to ensure you are in perfect condition for each operation?

Unger: In a natural way, I always remind myself that being a doctor is the foundation of my work. Through daily consultation and surgery, I can help my patients directly. Meeting them strengthens me, consulting with them is the source of my energy.

Ikeda: Jivaka was a celebrated doctor who lived in Shakyamuni's time. His name in Sanskrit means full of life or life-giver. I am sure that your own life-giving attitude sets your patient's minds at rest. You once said that the patient is the doctor's king and that all a doctor's knowledge and technical skill must be used for the patient's good. This is a very important assertion.

Unger: The doctor's duty and goal are to serve the patient as lord. The doctor must never regard the patient as a mere object or a means to an end. Each individual is a subject. I believe that patients' rights and position must be restored to their proper prominence.

Ikeda: People in leadership positions in any field, not just medicine, have a primary duty to serve the people. The goal of our SGI movement is to create an age in which ordinary people play the leading role and leaders serve them.

Patient's Right of Self-determination

Ikeda: As an issue related to medical ethics, I am interested to learn your thoughts on the patient's right of self-determination. Greater emphasis on it seems to have established patients' rights in therapy and equalized the patient–doctor relationship.

Unger: Yes, that is true. As those rights come into greater prominence, people speak out against their violation and against treating patients as things. We must act on the basis of a recognition that patients are subjects receiving treatment. It is completely appropriate for them to understand proposed therapies fully and to participate in judgments and decisions in connection with them.

Ikeda: Emphasis should be put on the rights of a patient to determine matters for him or herself. Still the decisions that must be made in modern therapy are highly diverse and demand consideration from many angles including those issues we have touched upon: surrogate parenthood, sale of organs for transplants, terminally ill patients' wish for euthanasia, and so on.

What are your thoughts on this expanded range of issues patients must decide?

Unger: I think patients have the right to choose a therapy. A good doctor will do his utmost to explain both the risk and benefit of the decisions they are making and motivate his patients to make the correct choices.

The doctor–patient dialogue is of central importance. Unfortunately, medicine today is losing the chance for the kind of dialogue a doctor can use to explain conditions and motivate patients for the sake of sound self-determination. My thirty years of experience have taught me that, given the right explanation and allowed to consider things completely, patients make the right choices. When people are ready to talk things over, many problems can be solved in ways satisfactory to both sides.

Ikeda: Today it is standard medical practice for doctors to explain therapy to patients thoroughly enough for them to understand and volitionally give what is called informed consent. This, of course, is premised on the ability of the patient to grant consent. How do you regard the situation when it involves minors?

Unger: When young children are involved, the issue of self-determination is somewhat different. In such cases, their parents and adults must be allowed to decide. A child is in no position to determine what is best. The same must be said of the severely intellectually and developmentally disabled. In such cases, intense individual examinations are essential. A doctor who feels obliged to respect the patient's right to life will be cautious not to deprive the patient of their right of self-determination any more than necessary and will motivate the patient to arrive at the best decision. As I have said, and as deserves repeating, dialogue and trust are the indispensable basis of medical treatment.

Medical Mishaps

Ikeda: I agree completely. Patients have the right to know and the right to be convinced about the therapy they are to undergo. Doctors have the obligation to explain the situation to them.

In Japan, an increasing frequency of mistakes on the part of doctors – mistaken surgery, wrong drug administration, and so on – has become a serious social problem. How do you think we can best prevent such mistakes?

Unger: Doctors are human beings liable to make mistakes. That is why it is essential that every error be exhaustively investigated to clarify its causes. Medicine today clings hard and fast to the defensive approach and starts dealing with each problem only after it has become serious. Moreover, doctors are burdened with hospital bureaucracy and mountains of paperwork. Computerization of information like patient records means that they spend more time in front of computers and much less with patients.

Ikeda: That is a worrying situation. Although undeniably medicine in general grows more complex and doctors' workload increases, the more medical technology advances, the more the doctor must diligently strive to be humane.

Unger: That is why I constantly insist that our hospital never become bureaucratic. A hospital bureaucracy works well only after patient numbers dwindle down to nothing. I do not want that to happen to our hospital. Our health system must be patient orientated.

Ikeda: You make an important point that applies to more than just hospitals. From their very origins, all kinds of organizations must decide for whom and to what purpose they exist. Observing you reminds me of the insistence in the Hippocratic Oath that knowledge and skills be used, not to exploit, but to serve humanity. This embodies the spirit of your academy.

Hippocrates also states that a physician who is a lover of wisdom is the equal of a god.[4] As a famous physician, the saver of many lives, and a great contributor to European intellectual growth, I pray that you will continue your activities with renewed fervour.

Unger: True value comes from the heart and can be transmitted from heart to heart. In this dialogue I have been deeply moved by your heartfelt words. I believe that on the basis of our many shared ideas we have created something very fruitful. I hope that you and I can continue to spread the spirit of tolerance throughout the world for the sake of peace.

Notes

Introduction

1   Cf. Leo Tolstoy, *The Kingdom of God Is Within You*, trans. Constance Garnett (Lincoln and London: University of Nebraska Press, 1984), p. 313.

2   Arnold Toynbee and Daisaku Ikeda, *Choose Life*, ed. Richard Gage (London: I.B.Tauris, 2007), p. 82.

3   http://www.scoop.co.nz/stories/WO0009/S00089/un2k-indone sian-president-abdurrahman-wahid.htm [retrieved on 4 December 2015].

4   *The Vimalakirti Sutra*, trans. Burton Watson (New York: Columbia University Press, 1997), p. 65.

5   Nichiren, *The Record of the Orally Transmitted Teachings* (Tokyo: Soka Gakkai, 2004), p. 138.

6   Aurelio Peccei and Daisaku Ikeda, *Before It Is Too Late*, ed. Richard Gage (London: I.B.Tauris, 2009), p. 92.

Chapter 1

1   http://www.euro-acad.eu/downloads/memorandas/charta_of_toler ance.pdf [retrieved on 5 December 2015].

2 Stefan Zweig, *The Right to Heresy: Castellio Against Calvin*, trans. Eden and Cedar Paul (Lexington, MA: Plunkett Lake Press, 2015), Kindle edition.

3 Ibid.

4 *The Edicts of King Asoka*, trans. Ven. S. Dhammika (Kandy, Sri Lanka: Buddhist Publication Society, 1993), p. 29.

5 Carl G. Jung, *The Undiscovered Self with Symbols and the Interpretation of Dreams*, trans. R.F.C. Hull (Princeton and Oxford: Princeton University Press, 1990), Kindle edition.

Chapter 2

1 Nichiren, *The Writings of Nichiren Daishonin* (Tokyo: Soka Gakkai, 1999), vol. I, p. 625.

2 *The Lotus Sutra and Its Opening and Closing Sutras*, trans. Burton Watson (Tokyo: Soka Gakkai, 2009), pp. 200–1.

3 *The Dhammapada*, trans. Eknath Easwaran (Tomales, CA: Nilgiri Press, 2007), p. 106.

4 *The Writings of Nichiren Daishonin*, vol. I, p. 937.

5 Nichiren, *The Writings of Nichiren Daishonin* (Tokyo: Soka Gakkai, 2006), vol. II, p. 759.

6 Johann Wolfgang von Goethe, *Hermann and Dorothea*, from Joseph Gostwick, *The Spirit of German Poetry: a series of translations from the German poets* (London: W. Smith, 1845), p. 46.

7 *The Record of the Orally Transmitted Teachings*, p. 90.

8 Translated from Japanese. Arnold Toynbee and Daisaku Ikeda, *Nijuisseiki eno taiwa* (Dialogue for the 21st Century) in *Ikeda Daisaku Zenshu* (The Complete Works of Daisaku Ikeda) (Tokyo: Seikyo Shimbun-sha, 1991), vol. 3, pp. 271–2.

9 *The Writings of Nichiren Daishonin*, vol. I, p. 1119.

10 Translated from Japanese. Richard Nicolaus Coudenhove-Kalergi and Daisaku Ikeda, *Bunmei: nishi to higashi* (Civilization, East and West) in *Ikeda Daisaku Zenshu* (The Complete Works of Daisaku Ikeda) (Tokyo: Seikyo Shimbun-sha, 2003), vol. 102, p. 94.

11 Richard Nicolaus Coudenhove-Kalergi, *Pan-Europe* (New York: A.A. Knopf, 1926), p. xvi.

12 Translated from Japanese. John Kenneth Galbraith and Daisaku Ikeda, *Ningen shugi no dai seiki o* (Dialogue for a greater century of humanism) (Tokyo: Ushio Shuppan-sha, 2005), pp. 199–200.

Chapter 3

1 *The Lotus Sutra and Its Opening and Closing Sutras*, p. 135.

2 Arnold Toynbee, *The World and the West* (London: Oxford University Press, 1953), p. 81.

3 Cf. Austregésilo de Athayde and Daisaku Ikeda, *Human Rights in the Twenty-first Century*, trans. Richard Gage (London: I.B.Tauris, 2009), p. 38.

4 Ibid., p. 51.

5 Ibid., p. 63.

6 Ibid., p. 56.

7 *The Writings of Nichiren Daishonin*, vol. I, p. 579.

8 See Hugo von Hofmannsthal, *Buch der Freunde* (Leipzig: Insel-Verlag, 1922), p. 5.

9 Ibid., p. 13.

10 *The Record of the Orally Transmitted Teachings*, p. 138.

11 *The Writings of Nichiren Daishonin*, vol. II, p. 931.

12 *The Writings of Nichiren Daishonin*, vol. I, p. 385.

13 http://www.euro-acad.eu/downloads/memorandas/charta_of_tolerance.pdf [retrieved on 5 December 2015].

14 Translated from German. Erich Neumann, *Zur Psychologie des Weiblichen* (Frankfurt am Main: Fischer Taschenbuch Verlag, 1983), p. 101.

15 Cf. David Winner, *Eleanor Roosevelt* (San Diego, CA: Blackbirch Press, 2003), p. 52.

16 Cf. Rabindranath Tagore, *The English Writings of Rabindranath Tagore*, ed. Sisir Kumar Das (New Delhi: Sahitya Akademi, 1996), vol. 2, p. 413.

17 Ibid., p. 416.

18 Ibid.

19 Mahatma Gandhi, *All Men Are Brothers* (New York: Continuum, 2000), p. 148.

20 Tsunesaburo Makiguchi, *A Geography of Human Life*, ed. Dayle Bethel (San Francisco: Caddo Gap Press, 2002), p. 286.

21 Translated from German. Richard Nikolaus Coudenhove-Kalergi, *Ethik und Hyperethik* (Leipzig: Verlag der Neue Geist, 1922), p. 115.

22 Cf. Arthur Stanley, *Stanley's Life of Thomas Arnold* (London: J. Murray, 1901), p. 94.

Chapter 4

1 *The Writings of Nichiren Daishonin*, vol. I, p. 4.

2 René Dubos, *So Human an Animal* (New York: Charles Scribner's Sons, 1968), pp. 203–4.

3 The Ten Worlds: (1) hell, (2) hungry spirits, (3) animals, (4) *asuras*, or egoistic pride (5) human beings, (6) heavenly beings, (7) voice-hearers, (8) cause-awakened ones, (9) bodhisattvas, and (10) Buddhas.

4 The Ten Factors of Life: appearance, nature, entity, power, influence, internal cause, relation, latent effect, manifest effect, and their consistency from beginning to end.

5 Three Realms of Existence: the realm of the five components, the realm of living beings, and the realm of the environment.

6 Translated from Japanese. Toynbee and Ikeda, *Nijuisseiki eno taiwa*, p. 241.

7 Gottfried W. Leibniz, *Discourse on Metaphysics and Other Essays*, trans. Daniel Garber and Roger Ariew (Indianapolis & Cambridge: Hackett Publishing Company, 1991), p. 39.

8 *The Group of Discourses (Sutta-Nipata)*, trans. K.R. Norman (Oxford: The Pali Text Society, 1995), vol. II, p. 17.

Chapter 5

1 *The Dhammapada*, p. 136.

2 *The Writings of Nichiren Daishonin*, vol. I, p. 851.

3 Mahatma Gandhi, *The Health Guide* (New York: The Crossing Press, 1978), p. 179.

4 *Hippocrates*, trans. W.H.S. Jones (Cambridge, MA: Harvard University Press, 2006), vol. II, p. 287.

Index

www.ingramcontent.com/pod-product-compliance
Lightning Source LLC
Chambersburg PA
CBHW061747270326
41928CB00011B/2407